ADVANCES IN LIPOPROTEIN RESEARCH

Edited by **Turgay İsbir**

Advances in Lipoprotein Research
http://dx.doi.org/10.5772/64122
Edited by Turgay İsbir

Contributors

Kwang-Min Kim, Tayhas Palmore, Chesa Gift Chauke, Jurgen Victor Seier, Zandisiwe Emilia Magwebu, Kaustubh Bora, Probodh Borah, C. Roger White, Samantha Giordano, Geeta Datta, Jesus Crespo, Nuria Zamora-Gonzalez, M.Dolores Calvo-Nieves, Cristina Andres-Ledesma, Alexander Orekhov, Nikita Nikiforov, Emile Zakiev, Alexandra Melnichenko

Published by InTech
Janeza Trdine 9, 51000 Rijeka, Croatia

Notice

Statements and opinions expressed in the chapters are these of the individual contributors and not necessarily those of the editors or publisher. No responsibility is accepted for the accuracy of information contained in the published chapters. The publisher assumes no responsibility for any damage or injury to persons or property arising out of the use of any materials, instructions, methods or ideas contained in the book.

Publishing Process Manager Ana Simcic
Technical Editor SPi Global
Cover InTech Design team

First published March, 2017
Printed in Croatia
Legal deposit, Croatia: National and University Library in Zagreb

Additional hard copies can be obtained from orders@intechopen.com

Advances in Lipoprotein Research , Edited by Turgay İsbir
p. cm.
Print ISBN 978-953-51-3047-5
Online ISBN 978-953-51-3048-2

Meet the editor

Prof. Dr. Turgay İsbir is the director of Medical Biology at Medical Faculty and the Department of Molecular Medicine in the Institute of Health Sciences of Yeditepe University. He worked at the Department of Pharmacology in the Brain Research Council of Edinburgh University and Department of Hematology of the Children's Hospital at the University of Pennsylvania. He also worked at the Department of Molecular Medicine in the Mayo Clinic. He was the director of İstanbul University, Aziz Sancar Institute of Experimental Medicine. He has been working at Yeditepe University since 2010. Currently, his research interests are focused mainly on the relationship between genetic polymorphisms and various types of cancers, diseases of lipid metabolism, and genetic tendencies of cardiovascular diseases. He has more than 235 published papers.

Contents

Preface

Lipoproteins have key roles in human growth and development, along with promoting, preventing, and/or participating in the pathogenesis or in the treatment of various diseases. This book presents a systematic and comprehensive review about the structure and metabolism of lipoproteins, particularly highlighting the crucial role of these molecules in the body and considering the interest of some lipids in health and disease.

This book is about lipoproteins, the structures for water-insoluble lipid transportation in the circulation. Lipoprotein particles not only take roles in a variety of normal physiological processes but also play essential roles in various pathological conditions. They are important structures that show varying patterns that correlate with the risk of fatal cardiovascular events associated with a high risk of atherosclerosis, while high HDL is correlated with reduced cardiovascular risk. Lipoproteins are now known to be important in other conditions like inflammatory situations, as well. It was assumed that the reader is familiar with the general aspects of lipoproteins.

The book is made up of four main sections: "Lipoproteins in Metabolism," containing the subtopics about metabolic regulations of lipoproteins; "Genetic Variations of Lipoprotein Metabolism," about the genetic polymorphisms affecting the lipoproteins; "Inflammatory Conditions and Lipoproteins," the relationship of lipoproteins and inflammation; and "Lipoproteins in Specific Diseases," roles of lipoproteins especially in brain diseases and cancers. This book edition aims to provide integrative approach and enables understanding of the lipoprotein metabolism. Distinguished international experts contribute six chapters on the genetic variations, plasma lipoprotein components, and molecular relations of lipoproteins with cognition and obesity.

Prof. Dr. Turgay İsbir, Ph.D.
Chairman, Department of Medical Biology
Yeditepe University, Faculty of Medicine
Director, Department of Molecular Medicine
Yeditepe University, Institute of Health Science
Istanbul, Turkey

Section 1

Lipoproteins in Metabolism

Chapter 1

Circulating Atherogenic Multiple-Modified Low-Density Lipoprotein: Pathophysiology and Clinical Applications

Nikita G. Nikiforov, Emile R. Zakiev, Alexandra A. Melnichenko and Alexander N. Orekhov

Additional information is available at the end of the chapter

http://dx.doi.org/10.5772/67055

Abstract

Low-density lipoprotein (LDL) circulating in human bloodstream is the source of lipids that accumulate in arterial intimal cells in atherosclerosis. *In-vitro*–modified LDL (acetylated, exposed to malondialdehyde, oxidized with transition metal ions, etc.) is atherogenic, that is, it causes accumulation of lipids in cultured cells. We have found that LDL circulating in the atherosclerosis patients' blood is atherogenic, while LDL from healthy donors is not. Atherogenic LDL was found to be desialylated. Moreover, only the desialylated subfraction of human LDL was atherogenic. Desialylated LDL is generally denser, smaller, and more electronegative than native LDL. Consequently, these LDL types are multiply modified, and according to our observations, desialylation is probably the principal and foremost cause of lipoprotein atherogenicity. It was found that desialylated LDL of coronary atherosclerosis patients was also oxidized. Complex formation further increases LDL atherogenicity, with LDL associates, immune complexes with antibodies recognizing modified LDL and complexes with extracellular matrix components being most atherogenic. We hypothesized that a nonlipid factor might be extracted from the blood serum using a column with immobilized LDL. This treatment not only allowed revealing the nonlipid factor of blood atherogenicity but also opened the prospect for reducing atherogenicity in patients.

Keywords: atherosclerosis, multiple-modified LDL, desialylated LDL, atherogenicity, circulating immune complexes, therapeutic approach

1. Introduction

Early stages of atherosclerosis development are characterized by abnormally high lipid accumulation in the arterial intima [1]. Formation of foam cells filled with lipids may be considered as the onset of the disease [2]. Low-density lipoprotein (LDL) circulating in human bloodstream is the origin of lipids that accumulate in the arterial intima cells [3]. However, intracellular cholesteryl ester accumulation could not be induced *in vitro* by native LDL [4]. On the other hand, *in vitro* –modified LDLs (acetylated, exposed to malondialdehyde, oxidized with transition metal ions, etc.) were demonstrated to cause lipid accumulation in cultured cells [5, 6]. Moreover, the question whether the modified LDL forms obtained *in vitro* fully correspond to the profile of modified LDL existing *in vivo* remains controversial. Therefore, the research community faces here a paradox: on one hand, a well-grounded opinion indicates LDL as the main source of lipid accumulation in the arterial wall, and on the other hand, native LDL failed to induce intracellular lipid accumulation in cultured cells. At the same time, *in vitro*–modified LDL was found to be atherogenic. However, detection of modified LDL in the bloodstream appeared to be challenging: acetylated LDL could not be found in the bloodstream, and the existence of oxidized LDL *in vivo* could not be demonstrated directly. Auto-antibodies against LDL modified by malondialdehyde, which is considered as a model of oxidized LDL, have been found in circulation [7]. It has to be kept in mind, however, that LDL conjugated with malondialdehyde (MDA-LDL) is a purely artificial modification, which cannot form in a living organism. Despite the fact that oxidized LDL has not been found in the bloodstream, the occurrence of antibodies against MDA-LDL is usually regarded as evidence of the existence of oxidized LDL *in vivo* [7].

2. Discovery of desialylated LDL in blood

In order to study modified LDL in atherosclerosis, we isolated LDL fraction from the blood of healthy subjects and atherosclerotic patients. We aimed to demonstrate that LDL from atherosclerotic patients can induce lipid accumulation in cultured cells. As a model, we used smooth muscle α-actin-positive cells, isolated from the intima of human aorta. These cells have been demonstrated to deposit lipids in atherosclerotic lesions *in situ* [8]. The method for isolation and cultivation of these cells has been previously established by our group [8]. After being cultured for 7 days, smooth muscle α-actin-positive cells (SMA(+) cells) originating from uninvolved intima of human aorta were subjected to a 24-hour incubation in Medium 199 supplemented with 10% lipoproteindepleted serum from a normal subject, as well as with LDL fraction with concentration 5–500 μg of apolipoprotein B (apo B)/ml. In the majority of experiments with the LDL samples isolated from normal subjects, there was no significant intracellular accumulation of phospholipids and neutral lipids [9]. By contrast, in the majority of experiments with LDL obtained from the plasma of coronary atherosclerosis patients, the intracellular levels of free cholesterol and triglycerides increased by 1.5 times and of the level of cholesteryl esters increased 1.5 to 5 times. Higher concentrations of LDL had no added effect on the intracellular lipid level. The results of the described experiments have demonstrated that LDL fractions isolated from the blood of patients with atherosclerosis, but not from normal subjects, induced deposition of lipids in human vascular cells. This feature of LDL was referred to as atherogenicity [10].

What is the possible explanation of LDL atherogenicity? We focused on comparing the properties of atherogenic LDL circulating in the patients' blood and nonatherogenic LDL from healthy donors. One of the major observations we made was the significantly (2 to 3 times) decreased sialic acid (N-acetylneuraminic acid) in LDL isolated from patients with coronary atherosclerosis [11]. Sialic acid is a terminal residue of asparagine-bound biantennary carbohydrate chains in LDL glycoconjugate moiety. In case of its removal, galactose becomes the terminal exposed residue. It is therefore possible to use *Ricinus communis* agglutinin (RCA120), which has a strong affinity to the terminal galactose to specifically isolate desialylated LDL [12]. We applied the total LDL preparation on a column containing CNBr-activated agarose-bound RCA120. LDL with sialylated carbohydrate chains passed freely through the column, while desialylated LDL bound to the lectin sorbent and could later be eluted with 5–50 mM galactose. This method allowed us extracting subfractions of both sialylated and desialylated LDL from the total LDL preparation isolated from the blood of patients. Desialylated LDL was found to be only a fraction of the total LDL pool circulating in patients' blood. Using the lectin affinity columns and lectinsorbent assay, we demonstrated that the ratio of desialylated LDL in blood of patients with coronary atherosclerosis was 20–60% of the total LDL level, while for normolipidemic subjects, desialylated LDL accounted for 5–15% [89]. The sialic acid content in desialylated LDL subfraction isolated by lectin chromatography was 2–3 times lower than that of sialylated LDL [12].

We next studied the atherogenic properties of desialylated LDL. Cultured SMA(+) cells, derived from the intima of human aorta and incubated with sialylated LDL subfraction, had unaltered intracellular contents of phospholipids and neutral lipids [12]. By contrast, cells incubated with desialylated LDL demonstrated a 1.5- to 2-fold increase in the contents of lipids and nonesterified cholesterol, as well as a 2- to 7-fold surge in the cholesteryl esters content. Therefore, only the desialylated subfraction of human LDL was found to be atherogenic. Normally, sialylated LDL had no atherogenic effect and could be regarded as native unmodified LDL. In summary, we have isolated a subfraction of naturally occurring desialylated LDL that was able to induce lipid deposition in human arterial subendothelial cells.

3. Trans-sialydase: the unknown LDL-modifying enzyme

More than 98% of the sialic acid cleaved from LDL is not present in the free form in the blood but is transferred to various protein acceptors [13]. Therefore, the enzyme responsible for desialylation of LDL works as a trans-sialidase.

We found that, apart from LDL, other lipoproteins, glycoproteins, and gangliosides are also affected by the trans-sialidase activity. Free sialic acid can be transferred to glycoproteins and sphingolipids of human serum. It can also be transferred to a protein or a lipid moiety of lipoprotein particles. Both lipoprotein fraction of human blood serum and lipoprotein-deficient serum had sialidase activity, as demonstrated by gel-filtration chromatography. Trans-sialidase activity was shown to be present in lipoproteins, as well as in a free form. The mechanism of trans-sialidase interaction with lipoproteins remains to be elucidated.

Using affinity chromatography, we succeeded in extracting a 65-kDa protein from lipoprotein deficient serum, which was a likely candidate to be the trans-sialidase [13]. The isolated enzyme was present in quantities from 20 to 200 μg/ml of human serum. The enzyme had three pH optima: 3.0, 5.0, and 7.0. The optimal pH spectrum indicated that the trans-sialidase would be active both in blood and in cellular organelles with low pH. Calcium and magnesium ions at millimolar concentrations could influence the enzyme activity *in vitro*. Thiol groups were found to be essential for normal enzyme functioning. Various blood proteins could serve as substrates for trans-sialidase activity. The enzyme successfully cleaved sialic acids from HDL, LDL, IDL, and VLDL particles. Trans-sialidase could also cleave sialic acid residues from glycoconjugates found in plasma, proteins (fetuin and transferrin), and gangliosides (GM3, GD3, GM1, GD1a, and GD1b). The rate of sialic acid transfer from these glycoconjugates was, however, much slower as compared to LDL. Among the sialylated LDL, VLDL, IDL, and HDL, the former has the highest affinity to the trans-sialidase. The mechanism trans-sialidase preference for LDL is unclear. It is possible that trans-sialidase activity is affected by the particle volume.

Importantly, isolated naturally occurring trans-sialidase was able to desialylate native LDL, which resulted in formation of desialylated LDL, which could induce cholesteryl ester accumulation in SMA(+) human aortic intimal cells [13]. This underscores the possible role of the enzyme in foam cell formation.

The role of plasma trans-sialidase remains to be established. Possible functions of trans-sialidase may include regulation of plasma proteins activities, cell-to-cell interactions, lifespan of glycoproteins, lipoproteins, and cells, etc [14].

Given its role in the formation of modified LDL, trans-sialidase activity may be an important component in the onset and progression of atherosclerosis. Trans-sialidase can also affect the interaction of lipoproteins with the arterial wall. Lipid accumulation induced by lipoproteins processed by trans-sialidase can be associated with the induction of proliferation and extracellular matrix synthesis. In conclusion, trans-sialidase may participate in all currently known cellular manifestations of atherosclerosis.

4. Physical properties of desialylated LDL

LDL is defined as a lipoprotein fraction with densities spanning from 1.019 to 1.063 g/l. Using ultracentrifugation and gradient gel electrophoresis, LDL particles can be segregated into four subfractions, including large, intermediate, small, and very small LDL [15].

4.1. Size and density

We first separated LDL particles based on their flotation rate using analytical ultracentrifugation [16].

We have determined densities of both native and desialylated LDL using gradient density ultracentrifugation [17]. Desialylated LDL fraction tended to have higher density than native LDL. Increased density was caused by lower amounts of phospholipids, free and esterified cholesteryl, and triglycerides.

Another method of LDL analysis, gradient gel electrophoresis, allows for separation of LDL subfractions by their electrophoretic mobility, which depends on the particle size and shape [18]. Gradient gel electrophoresis separation allows distinguishing 4 subclasses: large, intermediate, small, and very small LDL [19]. Correlation of LDL particle size and density is highly significant, as shown using ultracentrifugation, as well as gradient gel electrophoresis. However, these parameters are not always equal. Another method to analyze the weight and size of LDL particles is capillary gel electrophoresis is also used [20].

Another relatively new method for analysis of LDL is nuclear magnetic resonance (NMR). It is sometimes used for analyzing LDL subfractions in blood plasma, although the results obtained using this method cannot be compared directly with those obtained by ultracentrifugation of gel electrophoresis [21]. Other available methods of LDL analysis include high-performance liquid chromatography [22], dynamic light scattering [23], ion mobility analysis [24], and homogenous assay analysis [25].

Desialylated and native LDL particle size was estimated by our group using quasi-elastic laser scattering in a lipoprotein suspension followed by electrophoresis in polyacrylamide gel and scanning densitometry [26]. Native LDL particles from healthy subjects and atherosclerotic patients had sizes of 26.5 and 26.8 nm, respectively. Desialylated LDL of healthy subjects and atherosclerotic patients were 24.8 and 24.5 nm, respectively. The results of polyacrylamide gel electrophoresis were similar, with average diameters of native LDL being 26.3 and 26.2 nm for controls and for patients and those of desialylated LDL being 23.5 and 22.9 nm for controls and patients, respectively [17]. These results demonstrated that desialylated LDL had a reduced particle size in comparison with native LDL.

The origins of LDL subfractions remain unclear. According to Berneis, two types of precursors are secreted by the liver: triglyceride-poor apoB and triglyceride-rich apoB [27]. Triglyceride-poor lipoprotein gives rise to the large LDL subfraction while triglyceride-rich lipoprotein is a precursor for small dense LDL. This hypothesis explains the formation of small dense LDL from liver-secreted precursors and is supported by clinical results [27].

Genome-wide association studies have been used to search the factors affecting small dense LDL production. The available results indicate that small dense LDL metabolism is connected to genetic factors that may be considered as potential therapeutic targets for treatment of atherosclerosis [24].

Small dense LDL has a higher lifetime than large LDL, which is retrieved from the bloodstream through the LDL receptor pathway [28]. Small dense LDL tends to have lower levels of vitamins and antioxidatants than normal LDL. This means that small dense LDL is more oxidation prone than the larger forms of LDL [29].

It has been demonstrated that incubation of native LDL particles with atherosclerosis patients' blood plasma results in a significant decrease of the sialic acid contents [13]. Small dense

LDL particles have been shown to contain less sialic acid than larger LDL particles [30]. The increased ability of small dense LDL to form complexes with proteoglycans leads to the prolonged residence time of these particles in the subendothelial space or the arterial wall, where LDL may contribute to the development of atherosclerotic lesion [31]. In summary, LDL particle density reversely correlates with the particle size and sialic acid contents and directly correlates with atherogenicity.

4.2. Electronegativity

Desialylated LDL has been demonstrated to have a 1.2- to 1.4-times increased electrophoretic motility in comparison to native LDL [17]. Therefore, desialylated LDL has a lower charge than native LDL, that is, is more electronegative.

Agarose gel electrophoresis allows for specific isolation of electronegative LDL (LDL(−)). Isotachophoresis or ion exchange chromatography can be used as well [32]. The group of Avogaro was the first to discover and isolate atherogenic LDL(−) fraction [32] using ion-exchange chromatography.

More recent studies have revealed heterogeneity of LDL(−) particles, defining as many as five subclasses of LDL(−) [33]. The majority of electronegative subfractions correlated with cardiovascular (CV) risks, including, but not limited to hypercholesterolemia, smoking, myocardial infarction, and diabetes mellitus type II [34].

Several methods have been developed for isolation and analysis of LDL(−). Capillary isotachophoresis is another method used for LDL(−) extraction and analysis [35]. This technique allows for separation of LDL(−) from other LDL particles by its migration rate. Heparin precipitate LDL(−) was analyzed using capillary isotachophoresis [36]. Monoclonal antibodies allow for distinguishing LDL(−) by specific epitopes [37]. LDL(−) ELISA method is based on this technique and may prove to be useful in clinical practice [38].

It has been demonstrated that LDL(−) particles tend to aggregate [32], and the LDL particle aggregates have been shown to be atherogenic [39]. Gnarled structure of the lipoprotein was shown to be the principal cause of LDL(−) association [40]. The secondary structure of apoB in LDL(−) appears to be disturbed [70], with tryptophan residues abnormally exposed to the aqueous environment [41] and lysine residues having an altered ionization state [42]. Lipid moieties of LDL(−) particles also affect their surface tension/fluidity, rendering the particles more-aggregation prone [43].

Moreover, improper folding of apolipoprotein of LDL(−) particles affects its affinity to LDL receptors, which in turn leads to extended blood circulation times of LDL(−) [44]. On the other hand, the most electronegative fraction of LDL(−) is able to bind to lectin-like oxidized LDL receptor 1 (LOX-1) [45]. That subfraction of LDL(−) when added to cultured endothelial cells is able to increase the production of reactive oxygen species and to upregulate C-reactive protein levels via LOX-1 signaling pathway [46].

Therefore, LDL(−) is able to provoke pro-inflammatory and immune responses that contribute to the progression of atherogenesis. LDL(−) forms complexes with proteoglycans in the subendothelial intima where it resides for extended amounts of time. Subendothelial cells take up LDL(−) via scavenger receptors, which leads to saturation of their cytoplasm with lipid deposits and results in foam cell formation. Autoantibodies to LDL(−) can contribute to the development of atherosclerosis as well [47]. LDL(−) is cytotoxic to endothelial cells, inducing apoptosis and provoking production of inflammatory molecules such as IL-8, VCAM-1, and MCP-1 [47]. Therefore, LDL(−) was demonstrated to be pro-atherogenic and pro-inflammatory.

4.3. Similarity of desialylated LDL with small dense LDL and LDL(−)

As discussed above, small dense LDL and LDL(−) are the forms of modified LDL that have been detected in human blood plasma [48]. Our group has performed a series of experiments comparing the properties of LDL particles modified *in vivo*. In a study conducted in collaboration with the group of Avogaro (Italy), we have demonstrated that the more electronegative LDL corresponds by its properties to desialylated LDL [49]. Desialylated LDL subfraction also turned out to be more electronegative [50]. Therefore, it is likely that desialylated LDL and electronegative LDL subfractions are similar if not identical. We have found desialylated LDL to be smaller and denser as compared to native LDL. Simultaneously, LaBelle and co-authors have shown that sialic acid centent was reduced in small dense LDL [51]. Therefore, converging evidence demonstrates that all modified LDL subfractions isolated by different methods may be the same subfraction that underwent multiple modifications.

4.4. Which of the LDL modifications conveys atherogenicity?

Atherogenic LDL naturally present in the blood was found to be small, dense, and highly electronegative. Atherogenic LDL is also characterized by altered protein, lipid, and carbohydrate compositions. Consequently, these LDL particles can be referred to as multiply modified. To understand which modifications convey LDL atherogenicity, we have investigated the relationship between changes in chemical and physical parameters of LDL and its ability to induce lipid accumulation in SMA(+) cells of human aortic intima. A significant reverse correlation ($r = -0.66$, $p < 0.05$) between LDL atherogenicity and the sialic acid content was observed. By contrast, no correlation was observed between atherogenicity and the LDL particle size and charge, as well as with the levels of phospholipids and neutral lipids. Levels of lipophilic antioxidants, lipid peroxidation products, free lysine amino groups, and susceptibility of LDL to oxidation did not correlate with atherogenicity significantly [50]. It is therefore likely that desialylation is the principal cause of lipoprotein atherogenicity.

5. Is oxidized LDL just a myth?

According to the mainstream current opinion, oxidized LDL is the main trigger of atherosclerosis [52]. However, oxidized LDL had never been found in blood. One of the possible explanations that has been proposed to explain this discrepancy was that LDL oxidation takes place not in the bloodstream but in the arterial wall. The oxidized LDL theory is based on the following observations:

- Antibodies against LDL-oxidized *in vitro* were able to bind to substrates originating from atherosclerotic lesions, where LDL was found to be co-localized with oxidation products [53].
- Part of LDL isolated from atherosclerotic plaques corresponds by its properties to oxidized LDL [54].
- Autoantibodies recognizing malondialdehyde-LDL have been found in the blood [7].

Worth noting here is the fact that circulating autoantibodies have affinity not to the oxidized LDL but rather to MDA-LDL, a model of oxidized LDL. Interestingly, antibodies recognizing MDA-LDL were demonstrated to have an even higher affinity to desialylated LDL [55]. It is therefore possible that anti-LDL autoantibodies that primarily react with desialylated LDL also show cross-reactivity with MDA-LDL. This observation, together with other facts, challenges the concept of the oxidative modification of LDL being the principal *in vivo* modification that causes the onset and progression of atherosclerotic lesions. Other modified LDL species that have been found in the blood probably deserve more attention from the scientific community.

5.1. Circulating atherogenic desialylated LDL is oxidized

The degree of lipoprotein oxidation is estimated by measuring the contents of hydroperoxides or thiobarbituric acid-reactive substances (TBARS). These compounds are usually formed in course of lipid peroxidation. However, chemical instability and hydrophilic nature of these substances may cause their loss from LDL particles during lipoprotein isolation and purification stages. We have established a new technique to evaluate the degree of LDL oxidation based on the assumption that chemically active lipid derivatives formed in the process of peroxidation are able to covalently bind to apoprotein B and thus may serve as a marker of lipoperoxidation occurring *in vivo* in lipoprotein particle [56]. We have discovered sterol and phosphates covalently bound to apoB in delipidated preparations of LDL oxidized by copper ions, azo-initiators, sodium hypochlorite, or cultured cells. Newly extracted and isolated LDL from healthy individuals contained no apoB-lipid adducts. It has been revealed that contrary to other parameters used to estimate the degree of lipid peroxidation in LDL, the level of cholesterol covalently bound to apoB of copper-oxidized LDL rised monotonously during incubation [50]. Therefore, the level of apoB-bound cholesterol is a parameter that reflects the degree of LDL oxidation.

Native LDL and desialylated LDL isolated from healthy subjects had apoB-bound cholesterol levels of 0.25 ± 0.08 and 0.28 ± 0.05 mol/mol apoB, respectively. ApoB-bound cholesterol level in native LDL of atherosclerotic patients did not differ significantly from its level in native LDL of healthy individuals. The content of apoB-bound cholesterol in desialylated LDL of patients was 7 times higher than in native LDL. Therefore, we have shown that desialylated LDL of coronary atherosclerosis patients is oxidized.

5.2. Desialylated LDL is oxidation prone

Desialylated LDL contains 2- to 4-fold more oxysterols compared to native LDL [17], which indicates the increased susceptibility of desialylated LDL to oxidation.

In addition to a high degree of *in vivo* oxidation, desialylated LDL possesses a higher susceptibility to *in vitro* oxidation, which was evaluated using the duration of lag-phase upon oxidation by copper ions [57]. Average duration of lag-phase of native LDL isolated from atherosclerotic patients did not differ from that of native LDL taken from healthy individuals. The lag-phase of desialylated LDL of healthy subjects and patients was significantly shorter (3- and 6-fold, respectively) than that of native LDL, indicating a higher *in vitro* proneness to oxidation of desialylated LDL. It should be noted that proneness to oxidation of total LDL preparations from healthy subjects and patients positively correlates with the proportion of desialylated LDL in the lipoprotein preparation.

In an attempt to find out the causes of increased degrees of *in vivo* oxidation and proneness to oxidation of desialylated LDL, we estimated the contents of major fat-soluble antioxidants in lipoprotein particles, analyzed dependences among the levels of tocopherols and carotenoids, coenzyme-Q10, and the concentration of cholesterol bound to apoB and duration of lag-phase.

The levels of all major antioxidants, including coenzymeQ10, lycopene, α-and y-tocopherols, and β-carotene, were 1.5 to 2 times lower in desialylated LDL than in native LDL. The amount of cholesterol bound to apoB in desialylated LDL positively correlated with the amount of ubiquinone and showed a negative correlation with ubiquinol and β-carotene concentrations. At the same time, a positive correlation was found between the amount of cholesterol bound to apoB and the ubiquinol level in native LDL. The length of lag-phase for desialylated LDL was positively associated with α-tocopherol and β-carotene amounts and negatively associated with the ubiquinone content. On the other hand, proneness to oxidation of native LDL positively correlated with ubiquinone level.

Based on these observations, we hypothesized that a) the levels of examined lipophilic antioxidants in desialylated LDL are lower than in native lipoproteins, which leads to the high proneness of desialylated LDL to oxidation; b) coenzyme-Q10 might play a pro-oxidational role in native LDL; c) *in vivo* lipid peroxidation in desialylated LDL is enhanced by the increased proportion of oxidized form of coenzyme-Q10; and d) the severity of *in vivo* oxidation in desialylated LDL is associated with oxidation degree of ubiquinol and the amount of carotenoids loss.

6. Mechanisms increasing LDL atherogenicity

Based on the known rates of LDL uptake and degradation by the arterial wall cells, we estimate the time necessary for a normal intimal cell to become a foam cell, give 130 years . This estimation implies that there occurs no cholesterol efflux from the cell. The same estimation for desialylated LDL brings the result reduced to 15 years. However, according to angiographic and ultrasonographic data, atherosclerotic plaque can reduce the carotid artery lumen by one half within several weeks or months. Therefore, the actual rate of foam cell formation should be much higher than the estimated one, indicative of some processes that enhance the atherogenicity of desialylated LDL.

6.1. Self-association of LDL

It has been demonstrated that modified LDL particles are susceptible of self-association [58]. A positive correlation between atherogenicity of modified LDL and the degree of LDL association was demonstrated [59]. A dramatic increase of lipid accumulation rate by SMA(+) cells cultured from human aortic intima was observed upon incubation with lipoprotein associates. LDL associates removal from the incubation medium by filtration through filters with pore diameter 0.1 μm, completely eliminated any intracellular lipid accumulation. Therefore, association enhances LDL atherogenicity.

The absorption rate of associated LDL was 5–20 times higher than that of nonassociated LDL particles [39]. Latex beads (competiting phagocytic cargo) and cytochalasin B (inhibitor of phagocytosis) both inhibited the uptake of LDL associates [39]. It is therefore likely that LDL associates are absorbed via phagocytosis. The intracellular degradation rate of associated modified LDL apoB was 2–5 times slower than the rate of degradation of apoB of nonassociated particles. Therefore, high atherogenicity of lipoprotein associates is a result of enhanced absorption via phagocytosis and slow intracellular degradation rate.

6.2. LDL complexes with extracellular matrix

We have also demonstrated that LDL can form complexes with collagen, elastin, and proteoglycans of human aortic intima, as well as with cellular debris [60]. These complexes, once added cell culture, stimulated intracellular accumulation of lipids. Experiments with iodinated LDL have shown an increased absorption and diminished intracellular degradation rate of lipoprotein complexes, as compared to individual lipoprotein particles.

6.3. LDL-immune complexes

Multiply-modified lipoproteins are likely to be immunogenic. We succeeded in isolating circulating complexes containing LDL and anti-LDL autoantibodies from the blood of the majority of patients with coronary atherosclerosis [61].

We have observed a positive correlation between the levels of LDL-containing immune complexes in blood serum and the severity of coronary and extra-coronary atherosclerosis [62].

We have extracted LDL from circulating immune complexes by affinity chromatography on agarose with immobilized goat polyclonal antibodies against human LDL [63]. LDL from circulating immune complexes appeared to be desialylated, small, dense, more electronegative and with decreased contents of neutral lipids and phospholipids, as well as neutral saccharides. ApoB tertiary structure was also altered. Therefore, the LDL particles isolated from circulating immune complexes were similar if not identical to the desialylated LDL characterized previously.

We isolated antibodies to modified LDL from blood plasma of patients with coronary atherosclerosis [55]. These autoantibodies were identified as immunoglobulin G with an isoelectric point of about 8.5 (8.1–9.0), capable of interacting with the protein but not the lipid moiety of LDL. These autoantibodies were able to interact with native, glycosylated, acetylated, and oxidized LDL, showing the highest affinity for malondialdehyde-treated LDL, desialylated LDL, and LDL isolated from patients with coronary atherosclerosis.

Autoantibodies bound to native LDL forming complexes that could induce lipid aggregation in SMA(+) cells cultured from uninvolved intima of human aorta. Moreover, autoantibodies enhanced the atherogenic properties of desialylated LDL via complex formation [55]. It was found that C1q complement component and fibronectin could bind to the LDL-antibody complexes leading to a more pronounced lipid aggregation in SMA(+) human aortic intimal cells. C1q complement component is produced by dendritic cells in the spleen, where C1q is binds to immune complexes [64]. Antigen-presenting dendritic cells are also present in atherosclerotic plaques [65]. Moreover, dendritic cells expressing C1q have been found in atherosclerotic plaques [66]. C1q was also expressed in macrophages, foam cells, and in neovascular endothelial cells [66]. Thus, C1q expression might be an important feature of cells located in the vessel wall of atherosclerotic lesions, causing them to capture and retain immune complexes [66].

In vitro interaction of mouse peritoneal and human pericardial macrophages with immune complexes isolated from blood serum of ischemic heart disease patients led to the transformation of macrophages into foam cells [67]. Macrophages incubated with immune complexes for 3 days acquired cytoplasmic lipid vacuoles, and the cisterns of endoplasmic reticulum (ER) in these cells were dramatically enlarged and filled with lipids. The accumulation of lipids within ER cisterns in macrophages may be accompanied by ER stress, which also plays a role in the development of atherosclerosis [68].

From Alexander N. Orekhov, Alexandra A. Melnichenko, and Igor A. Sobenin, "Approach to Reduction of Blood Atherogenicity," Oxidative Medicine and Cellular Longevity, Vol. 2014, Article ID 738679, 8 pages, 2014. doi:10.1155/2014/738679.

7. Revealing a nonlipid factor of blood atherogenicity

We hypothesized that a nonlipid factor could be extracted from the serum using a column with immobilized LDL. We applied atherogenic serum, which has previously been shown to induce a nearly 5-fold increase of cholesterol content in cultured cells, to a column with LDL covalently bound to agarose. We found that the eluted serum lost its atherogenicity, i.e. it failed to induce a statistically significant lipid accumulation in cultured cells (**Figure 1**, from Ref. [69]). The substances retained on the column were eluted with glycine buffer and mixed with the sera samples that were previously treated by passing though the column, which resulted in the recovery of serum atherogenicity up to the initial level (**Figure 1**). It is therefore likely that patients' blood serum contains unknown atherogenic factors that can be absorbed on immobilized LDL.

We next used this method was to reduce atherogenicity of the blood of patients by extracorporeal perfusion. Four male volunteers aged 46–59 years with CHD, normal cholesterol levels, functional class II–III angina pectoris, and angiographically documented stenosis of 2 to 3 coronary arteries have agreed to take part in the study [69]. Three individuals were smokers, and one had mild arterial hypertension. A pronounced decrease of plasma atherogenicity was registered after a 2-hour extracorporeal perfusion through a column with autologous LDL (**Figure 2(a)**, from Ref. [69]).

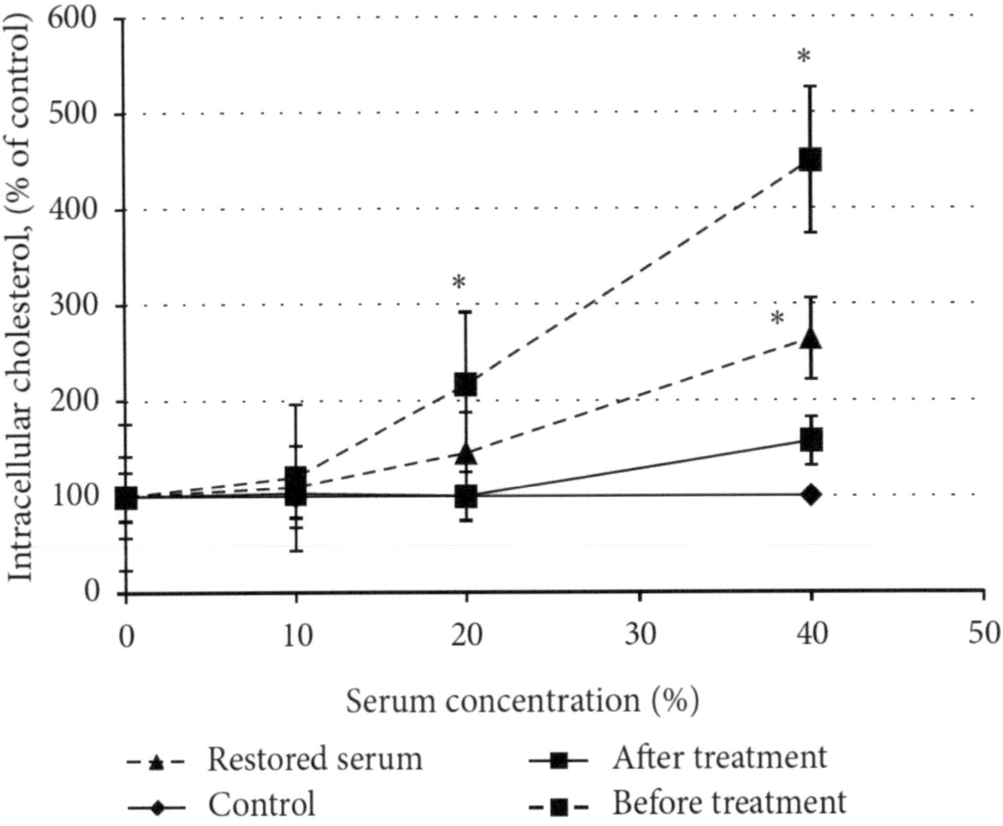

Figure 1. Elimination of serum atherogenicity with LDL-agarose column. Five milliliters of the serum were passed through the LDL sepharose column at a flow rate of 1 ml/min for 30 min. The sorbent was then eluted with 2-ml glycine buffer (pH 2.7), and the eluate was dialyzed against a 2000-fold excessive volume of medium 199 for 24 hours at 4°C. The cells were cultured in the presence of the initial or treated serum and with the proper volume of the dialyzed eluate.

The analysis of serum atherogenicity demonstrated that, in all four cases, it was reduced to a near-zero level 24 hours after the procedure and then gradually reappeared, reaching a significant level within 1 week. Repeated procedure resulted in a pronounced decrease of serum atheorgenicity, with the 2nd and the 3rd procedures reducing it for prolonged periods sufficient for reducing the frequency of the treatment. When applied once every 2 to 3 weeks, the procedure provided low levels of plasma atherogenicity for long periods (**Figure 2(b)**, from Ref. [69]). The procedure has been applied twice a month in one patient for 9 months and in another patient for more than 7 months. Each patient was examined taking into account the general state of health, number of angina pectoris episodes, the amount of medicine (nitrates) taken, and capacity for exercise. Bicycle test, 24-hour Holter ECG monitoring, and control of hematological and biochemical parameters have been performed every 3 months. During this trial, the patients have felt better, moved from functional class III to II (according to Canadian classification), and endured greater physical loads in the bicycle test [69]. Arterial blood pressure of patient 1 stabilized and reached a nearly normal level. Both patients have noted a heightened sexual activity and have associated this with reduced angina pectoris [69]. The repeated angiograms have been assessed after 20–25 months of treatment. There were no new stenoses, 50% stenoses have progressed, 25% regressed, and 25% remained unchanged. These observations suggest an improved disease progression in comparison to the normal course of coronary atherosclerosis [70].

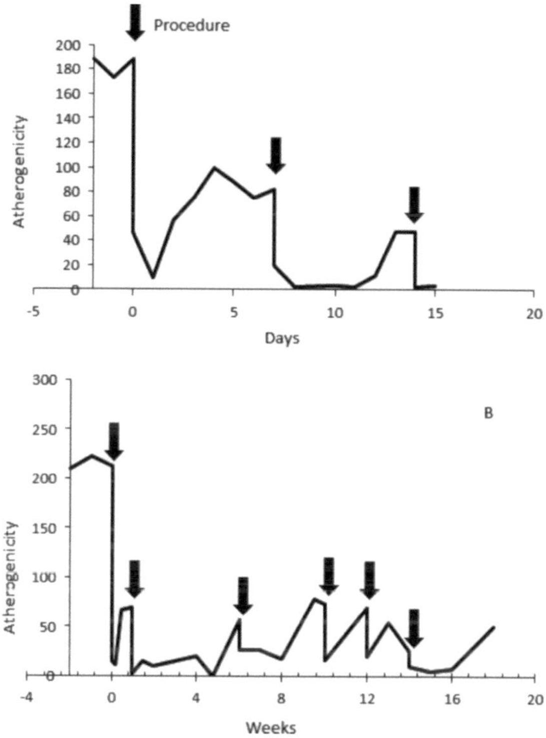

Figure 2. Monitoring of atherogenicity. The patient's plasma was subjected to 2-hour extracorporeal perfusion through a column with 200 mL of the sorbent; the flow rate was 30 ml/min. The total plasma volume of 2–3 liters was perfused through the column during the procedure. Blood serum atherogenicity after 3 procedures was assessed daily ((a), patient 3) and once or twice a week afterwards ((b), patient 1). Ordinate (atherogenicity), percent of cholesterol accumulation in the cells cultured in the presence of the serum from the CHD patient.

8. Multiple-modified LDL: which type of LDL modification occurs early in the blood?

What are the mechanisms of multiple modifications of LDL? Do they take place in the blood plasma? A round the clock exposure of LDL to hepatocytes, intact endotheliocytes, smooth muscle cells, macrophages, or cell homogenates has not affected properties of native LDL [13].

After incubation for 24 hours at 37°C with whole blood or plasma taken from patients with coronary atherosclerosis, the sialic acid content of LDL became 2 times lower than that of LDL incubated with whole blood or plasma obtained from healthy individuals. Incubation with red and white blood cells had no effect on the sialic acid content. This points out that LDL modification takes place in the blood plasma [13].

A detailed analysis of LDL modification processes has been performed by our group [13]. Native LDL was extracted from the blood plasma using ultracentrifugation followed by lectin chromatography. Serum was cleared from apoB-containing lipoproteins by defibrination of the remaining LDL-deficient plasma. Afterwards, LDL and serum were reconstituted in the same proportion in the original plasma and incubated for different time points at 37°C.

After incubation, LDL was re-isolated by ultracentrifugation. The described method allowed for elimination of the effects of LDL originating from VLDL and IDL in the process of incubation. After 1 hour of incubation of native LDL with autologous plasma samples, a sharp decrease of sialic acid content was observed. At the same time, desialylated LDL concentration increased, as determined by lectin-sorbent chromatography (**Table 1**, from Ref. [13]). In parallel to the decrease of the sialic acid content, LDL acquired capability to induce a pronounced accumulation of cholesterol in SMA(+) cells cultured from unaffected human aortic intima. This could be registered as early as after 3 hours of incubation. After 6 hours of incubation with plasma, a steady decrease of phospholipid and neutral lipid contents, as well as LDL particle size could be observed.

1 h	3 h	6 h	12 h	24 h	36 h	48 h
↓Sialic acid ↑% of desialylated LDL	↑ Atherogenicity	↓Free cholesterol	↓Size ↓Cholesteryl esters ↓Phospholipids	↓Triglycerides	↑ Electronegativity	↑Apo B-bound cholesterol ↑Susceptibility to oxidation ↑Fluorescence ↓Vitamin E

Table 1. The outline of LDL modification.

After 36 hours of incubation, negative charge of lipoprotein particles became obvious. Longer incubation times (48 and 72 hours) led to a loss of α-tocopherol and to an increase of LDL susceptibility to oxidation, as well as to aggregation of cholesterol covalently bound to apoB. Degradation of apoB was also registered at this point.

It can be concluded that desialylation is likely to be the primary and most important LDL modification that conveys its atherogenicity. Other known modifications may further increase the LDL atherogenicity.

9. Conclusion

We have obtained an LDL subfraction that was able to induce accumulation of lipids, primarily cholesteryl esters, in cultured SMA(+) cells. This helped to reconcile the facts that native LDL is not atherogenic and *in vitro*–modified LDL not present in circulation.

We have shown that atherogenic LDL is characterized by numerous alterations of carbohydrate, protein, and lipid moieties, and can therefore be termed multiple-modified LDL. Multiple modifications of LDL occur in human blood plasma. It was shown that circulating multiple-modified LDL loses the affinity for the B,E-receptor and acquires the ability to interact with a number of other cellular membrane receptors and proteoglycans. The enhanced cellular uptake of desialylated LDL, low degradation rate of apolipoprotein and cholesteryl esters, as well as stimulation of re-esterification of free cholesterol, cause the intracellular accumulation of intracellular-esterified cholesterol.

The formation of LDL-containing large complexes (associates, immune complexes, and complexes with the extracellular matrix components) can stimulate lipid accumulation in intimal smooth muscle cells. In addition to cholesteryl ester accumulation, desialylated LDL stimulates cell proliferation and synthesis of the connective tissue matrix.

Therefore, we have been able to obtain and describe naturally occurring multiple-modified LDL capable of provoking all atherosclerotic manifestations at the cellular level.

Immune complexes, consisting of LDL and autoantibodies, have been discovered in the human blood stream circulation [71]. Amount of LDL-containing circulating immune complexes was directly correlated with the severity of atherosclerosis [71]. We hypothesize that anti-LDL autoantibodies and circulating immune complexes containing LDL can be the factors that convey blood atherogenicity. Although the anti-LDL cannot be proven the only atherogenic factor adsorbed on the column with immobilized LDL, the substances binding to LDL should be thoroughly studied. Columns with immobilized LDL allowed not only distinguishing and collect nonlipid factors of atherogenicity, but also opening a prospect for reducing atherogenicity in patients.

Acknowledgements

We thank Ekaterina Ivanova for assistance in editing the manuscript. This work was supported by the Russian Foundation for Basic Research (Grant # 15-04-09279).

Author details

Nikita G. Nikiforov[1,2*], Emile R. Zakiev[1,3], Alexandra A. Melnichenko[1] and Alexander N. Orekhov[1,4]

*Address all correspondence to: nikiforov.mipt@googlemail.com

1 Laboratory of Angiopathology, Institute of General Pathology and Pathophysiology, Moscow, Russia

2 Laboratory of Medical Genetics, A. N. Myasnikov Institute of Clinical Cardiology, Russian Cardiology Research and Production Complex, Ministry of Healthcare, Moscow, Russia

3 INSERM UMR_S 1166, University of Pierre and Marie Curie, Paris, France

4 Institute for Atherosclerosis Research, Skolkovo Innovative Center, Moscow, Russia

References

[1] Reynolds T. Cholesteryl ester storage disease: a rare and possibly treatable cause of premature vascular disease and cirrhosis. J Clin Pathol. 2013;66(11):918-923.

[2] Uitz E, Bahadori B, McCarty MF, *et al.* Practical strategies for modulating foam cell formation and behavior. World J Clin Cases. 2014;2(10):497-506.

[3] Goldstein JL, Anderson RG, Brown MS. Coated pits, coated vesicles, and receptor-mediated endocytosis. Nature. 1979;279(5715):679-685.

[4] Brown MS, Goldstein JL, Krieger M, *et al.* Reversible accumulation of cholesteryl esters in macrophages incubated with acetylated lipoproteins. J Cell Biol. 1979;82(3):597-613.

[5] Fogelman AM, Shechter I, Seager J, *et al.* Malondialdehyde alteration of low density lipoproteins leads to cholesteryl ester accumulation in human monocyte-macrophages. Proc Natl Acad Sci U S A. 1980;77(4):2214-2228.

[6] Steinbrecher UP, Parthasarathy S, Leake DS, *et al.* Modification of low density lipoprotein by endothelial cells involves lipid peroxidation and degradation of low density lipoprotein phospholipids. Proc Natl Acad Sci U S A. 1984;81(12):3883-3897.

[7] Palinski W, Rosenfeld ME, Yla-Herttuala S, *et al.* Low density lipoprotein undergoes oxidative modification *in vivo*. Proc Natl Acad Sci U S A. 1989;86(4):1372-1376.

[8] Orekhov AN, Tertov VV, Novikov ID, *et al.* Lipids in cells of atherosclerotic and uninvolved human aorta. I. Lipid composition of aortic tissue and enzyme-isolated and cultured cells. Exp Mol Pathol. 1985;42(1):117-137.

[9] Orekhov AN, Tertov VV, Pokrovsky SN, *et al.* Blood serum atherogenicity associated with coronary atherosclerosis. Evidence for nonlipid factor providing atherogenicity of low-density lipoproteins and an approach to its elimination. Circ Res. 1988;62(3):421-429.

http://dx.doi.org/10.5772/67055

[10] Chazov EI, Tertov VV, Orekhov AN, *et al.* Atherogenicity of blood serum from patients with coronary heart disease. Lancet. 1986;2(8507):595-598.

[11] Orekhov AN, Tertov VV, Mukhin DN. Desialylated low density lipoprotein—naturally occurring modified lipoprotein with atherogenic potency. Atherosclerosis. 1991;86(2–3):153-161.

[12] Tertov VV, Sobenin IA, Tonevitsky AG, *et al.* Isolation of atherogenic modified (desialylated) low density lipoprotein from blood of atherosclerotic patients: separation from native lipoprotein by affinity chromatography. Biochem Biophys Res Commun. 1990;167(3):1122-1127.

[13] Tertov VV, Kaplun VV, Sobenin IA, *et al.* Low-density lipoprotein modification occurring in human plasma possible mechanism of *in vivo* lipoprotein desialylation as a primary step of atherogenic modification. Atherosclerosis. 1998;138(1):183-195.

[14] Monti E, Bonten E, D'Azzo A, *et al.* Sialidases in vertebrates: a family of enzymes ailored for several cell functions. Adv Carbohydr Chem Biochem. 2010;64:403-479.

[15] Mills EJ, Rachlis B, Wu P, *et al.* Primary prevention of cardiovascular mortality and events with statin treatments: a network meta-analysis involving more than 65,000 patients. J Am Coll Cardiol. 2008;52(22):1769-1781.

[16] Krauss RM, Lindgren FT, Ray RM. Interrelationships among subgroups of serum lipoproteins in normal human subjects. Clin Chim Acta. 1980;104(3):275-290.

[17] Tertov VV, Orekhov AN. Metabolism of native and naturally occurring multiple modified low density lipoprotein in smooth muscle cells of human aortic intima. Exp Mol Pathol. 1997;64(3):127-145.

[18] Ensign W, Hill N, Heward CB. Disparate LDL phenotypic classification among 4 different methods assessing LDL particle characteristics. Clin Chem. 2006;52(9):1722-1727.

[19] Williams PT, Vranizan KM, Krauss RM. Correlations of plasma lipoproteins with LDL subfractions by particle size in men and women. J Lipid Res. 1992;33(5):765-774.

[20] Hoefner DM, Hodel SD, O'Brien JF, *et al.* Development of a rapid, quantitative method for LDL subfractionation with use of the Quantimetrix Lipoprint LDL System. Clin Chem. 2001;47(2):266-274.

[21] Witte DR, Taskinen MR, Perttunen-Nio H, *et al.* Study of agreement between LDL size as measured by nuclear magnetic resonance and gradient gel electrophoresis. J Lipid Res. 2004;45(6):1069-1076.

[22] Okazaki M, Usui S, Ishigami M, *et al.* Identification of unique lipoprotein subclasses for visceral obesity by component analysis of cholesterol profile in high-performance liquid chromatography. Arterioscler Thromb Vasc Biol. 2005;25(3):578-584.

[23] Sakurai T, Trirongjitmoah S, Nishibata Y, *et al.* Measurement of lipoprotein particle sizes using dynamic light scattering. Ann Clin Biochem. 2010;47(Pt 5):476-481.

[24] Musunuru K, Orho-Melander M, Caulfield MP, *et al.* Ion mobility analysis of lipoprotein subfractions identifies three independent axes of cardiovascular risk. Arterioscler Thromb Vasc Biol. 2009;29(11):1975-1980.

[25] Hirano T, Ito Y, Saegusa H, *et al.* A novel and simple method for quantification of small, dense LDL. J Lipid Res. 2003;44(11):2193-2201.

[26] Orekhov AN, Tertov VV, Kudryashov SA, *et al.* Triggerlike stimulation of cholesterol accumulation and DNA and extracellular matrix synthesis induced by atherogenic serum or low density lipoprotein in cultured cells. Circ Res. 1990;66(2):311-320.

[27] Berneis KK, Krauss RM. Metabolic origins and clinical significance of LDL heterogeneity. J Lipid Res. 2002;43(9):1363-1379.

[28] Packard C, Caslake M, Shepherd J. The role of small, dense low density lipoprotein (LDL): a new look. Int J Cardiol. 2000;74 Suppl 1:S17-S22.

[29] Tribble DL, Rizzo M, Chait A, *et al.* Enhanced oxidative susceptibility and reduced antioxidant content of metabolic precursors of small, dense low-density lipoproteins. Am J Med. 2001;110(2):103-110.

[30] La Belle M, Krauss RM. Differences in carbohydrate content of low density lipoproteins associated with low density lipoprotein subclass patterns. J Lipid Res. 1990;31(9):1577-1588.

[31] Anber V, Griffin BA, McConnell M, *et al.* Influence of plasma lipid and LDL-subfraction profile on the interaction between low density lipoprotein with human arterial wall proteoglycans. Atherosclerosis. 1996;124(2):261-271.

[32] Avogaro P, Bon GB, Cazzolato G. Presence of a modified low density lipoprotein in humans. Arteriosclerosis. 1988;8(1):79-87.

[33] Yang CY, Raya JL, Chen HH, *et al.* Isolation, characterization, and functional assessment of oxidatively modified subfractions of circulating low-density lipoproteins. Arterioscler Thromb Vasc Biol. 2003;23(6):1083-1090.

[34] Tang D, Lu J, Walterscheid JP, *et al.* Electronegative LDL circulating in smokers impairs endothelial progenitor cell differentiation by inhibiting Akt phosphorylation via LOX-1. J Lipid Res. 2008;49(1):33-47.

[35] Schmitz G, Mollers C, Richter V. Analytical capillary isotachophoresis of human serum lipoproteins. Electrophoresis. 1997;18(10):1807-1813.

[36] Zhang B, Kaneshi T, Ohta T, *et al.* Relation between insulin resistance and fast-migrating LDL subfraction as characterized by capillary isotachophoresis. J Lipid Res. 2005;46(10):2265-2277.

[37] Damasceno NR, Sevanian A, Apolinario E, *et al.* Detection of electronegative low density lipoprotein (LDL-) in plasma and atherosclerotic lesions by monoclonal antibody-based immunoassays. Clin Biochem. 2006;39(1):28-38.

[38] Faulin Tdo E, de Sena-Evangelista KC, Pacheco DB, *et al*. Development of immunoassays for anti-electronegative LDL autoantibodies and immune complexes. Clin Chim Acta. 2012;413(1–2):291-297.

[39] Tertov VV, Orekhov AN, Sobenin IA, *et al*. Three types of naturally occurring modified lipoproteins induce intracellular lipid accumulation due to lipoprotein aggregation. Circ Res. 1992;71(1):218-228.

[40] Brunelli R, Balogh G, Costa G, *et al*. Estradiol binding prevents ApoB-100 misfolding in electronegative LDL(−). Biochemistry. 2010;49(34):7297-7302.

[41] Parasassi T, Bittolo-Bon G, Brunelli R, *et al*. Loss of apoB-100 secondary structure and conformation in hydroperoxide rich, electronegative LDL(−). Free Radic Biol Med. 2001;31(1):82-89.

[42] Blanco FJ, Villegas S, Benitez S, *et al*. 2D-NMR reveals different populations of exposed lysine residues in the apoB-100 protein of electronegative and electropositive fractions of LDL particles. J Lipid Res. 2010;51(6):1560-1565.

[43] Benitez S, Sanchez-Quesada JL, Lucero L, *et al*. Changes in low-density lipoprotein electronegativity and oxidizability after aerobic exercise are related to the increase in associated non-esterified fatty acids. Atherosclerosis. 2002;160(1):223-232.

[44] Benitez S, Villegas V, Bancells C, *et al*. Impaired binding affinity of electronegative low-density lipoprotein (LDL) to the LDL receptor is related to nonesterified fatty acids and lysophosphatidylcholine content. Biochemistry. 2004;43(50):15863-15872.

[45] Lu J, Yang JH, Burns AR, *et al*. Mediation of electronegative low-density lipoprotein signaling by LOX-1: a possible mechanism of endothelial apoptosis. Circ Res. 2009;104(5):619-627.

[46] Chu CS, Wang YC, Lu LS, *et al*. Electronegative low-density lipoprotein increases C-reactive protein expression in vascular endothelial cells through the LOX-1 receptor. PLoS One. 2013;8(8):e70533.

[47] Greco G, Balogh G, Brunelli R, *et al*. Generation in human plasma of misfolded, aggregation-prone electronegative low density lipoprotein. Biophys J. 2009;97(2):628-635.

[48] Krauss RM, Burke DJ. Identification of multiple subclasses of plasma low density lipoproteins in normal humans. J Lipid Res. 1982;23(1):97-104.

[49] Tertov VV, Bittolo-Bon G, Sobenin IA, *et al*. Naturally occurring modified low density lipoproteins are similar if not identical: more electronegative and desialylated lipoprotein subfractions. Exp Mol Pathol. 1995;62(3):166-172.

[50] Tertov VV, Sobenin IA, Orekhov AN. Characterization of desialylated low-density lipoproteins which cause intracellular lipid accumulation. Int J Tissue React. 1992;14(4):155-162.

[51] Tertov VV, Sobenin IA, Orekhov AN. Similarity between naturally occurring modified desialylated, electronegative and aortic low density lipoprotein. Free Radic Res. 1996;25(4):313-319.

[52] Arai H. Oxidative modification of lipoproteins. Subcell Biochem. 2014;77:103-114.

[53] Fukuchi M, Watanabe J, Kumagai K, *et al.* Normal and oxidized low density lipoproteins accumulate deep in physiologically thickened intima of human coronary arteries. Lab Invest. 2002;82(10):1437-1447.

[54] Yla-Herttuala S, Palinski W, Rosenfeld ME, *et al.* Lipoproteins in normal and atherosclerotic aorta. Eur Heart J. 1990;11 Suppl E:88-99.

[55] Orekhov AN, Tertov VV, Kabakov AE, *et al.* Autoantibodies against modified low density lipoprotein. Nonlipid factor of blood plasma that stimulates foam cell formation. Arterioscler Thromb. 1991;11(2):316-326.

[56] Tertov VV, Kaplun VV, Dvoryantsev SN, *et al.* Apolipoprotein B-bound lipids as a marker for evaluation of low density lipoprotein oxidation *in vivo*. Biochem Biophys Res Commun. 1995;214(2):608-613.

[57] Esterbauer H, Striegl G, Puhl H, *et al.* Continuous monitoring of *in vitro* oxidation of human low density lipoprotein. Free Radic Res Commun. 1989;6(1):67-75.

[58] Tertov VV, Sobenin IA, Gabbasov ZA, *et al.* Multiple-modified desialylated low density lipoproteins that cause intracellular lipid accumulation. Isolation, fractionation and characterization. Lab Invest. 1992;67(5):665-675.

[59] Tertov VV, Sobenin IA, Gabbasov ZA, *et al.* Lipoprotein aggregation as an essential condition of intracellular lipid accumulation caused by modified low density lipoproteins. Biochem Biophys Res Commun. 1989;163(1):489-494.

[60] Orekhov AN, Tertov VV, Mukhin DN, *et al.* Association of low-density lipoprotein with particulate connective tissue matrix components enhances cholesterol accumulation in cultured subendothelial cells of human aorta. Biochim Biophys Acta. 1987;928(3):251-258.

[61] Sobenin IA, Salonen JT, Zhelankin AV, *et al.* Low density lipoprotein-containing circulating immune complexes: role in atherosclerosis and diagnostic value. Biomed Res Int. 2014;2014:205697.

[62] Kacharava AG, Tertov VV, Orekhov AN. Autoantibodies against low-density lipoprotein and atherogenic potential of blood. Ann Med. 1993;25(6):551-555.

[63] Tertov VV, Sobenin IA, Orekhov AN, *et al.* Characteristics of low density lipoprotein isolated from circulating immune complexes. Atherosclerosis. 1996;122(2):191-199.

[64] Schwaeble W, Schafer MK, Petry F, *et al.* Follicular dendritic cells, interdigitating cells, and cells of the monocyte-macrophage lineage are the C1q-producing sources in the spleen. Identification of specific cell types by *in situ* hybridization and immunohistochemical analysis. J Immunol. 1995;155(10):4971-4978.

[65] Bobryshev YV, Lord RS. Ultrastructural recognition of cells with dendritic cell morphology in human aortic intima. Contacting interactions of vascular dendritic cells in athero-resistant and athero-prone areas of the normal aorta. Arch Histol Cytol. 1995;58(3):307-322.

[66] Cao W, Bobryshev YV, Lord RS, *et al*. Dendritic cells in the arterial wall express C1q: potential significance in atherogenesis. Cardiovasc Res. 2003;60(1):175-186.

[67] Klimov AN, Denisenko AD, Vinogradov AG, *et al*. Accumulation of cholesteryl esters in macrophages incubated with human lipoprotein-antibody autoimmune complex. Atherosclerosis. 1988;74(1–2):41-46.

[68] Chistiakov DA, Sobenin IA, Orekhov AN, *et al*. Role of endoplasmic reticulum stress in atherosclerosis and diabetic macrovascular complications. Biomed Res Int. 2014;2014:610140.

[69] Orekhov AN, Melnichenko AA, Sobenin IA. Approach to reduction of blood atherogenicity. Oxid Med Cell Longev. 2014;2014:738679.

[70] Loukas M, Sharma A, Blaak C, *et al*. The clinical anatomy of the coronary arteries. J Cardiovasc Transl Res. 2013;6(2):197-207.

[71] Sobenin IA, Karagodin VP, Melnichenko AC, *et al*. Diagnostic and prognostic value of low density lipoprotein-containing circulating immune complexes in atherosclerosis. J Clin Immunol. 2013;33(2):489-495.

Chapter 2

Apolipoprotein D

Jesús Crespo-Sanjuán, Nuria Zamora-Gonzalez,
María Dolores Calvo-Nieves and
Cristina Andres-Ledesma

Additional information is available at the end of the chapter

http://dx.doi.org/10.5772/66626

Abstract

Apolipoprotein D (ApoD) is an extracellular glycoprotein of the lipocalin protein family, involved in different functions such as immune response, cell proliferation regulation, chemoreception, retinoid metabolism, axon growth, and proteolysis regulation. This lipocalin is expressed predominantly in the nervous system (NS), both prenatally (vascular pericytes) and postnatally (glia and neurons) and in adulthood. It is also expressed in other tissues and is carried by high density lipoprotein (HDL) in plasma, so it could interfere in cholesterol and other lipids regulation. ApoD increases considerably in systemic apocrine gland tumors and also in some primary brain tumors. Although the specific biological role of ApoD is unknown, the presence of ApoD in tumors appears to be a prognostic factor in their evolution. Regarding the NS, increased ApoD expression observed in many neurodegenerative diseases could be used to make an early diagnosis thereof.

Keywords: oxidative stress, apolipoprotein, nervous system, colorectal cancer

1. Introduction

Apolipoprotein D (ApoD) is an extracellular glycoprotein of the lipocalin protein family, involved in different functions such as immune response, cell proliferation regulation, chemoreception, retinoid metabolism, axon growth, and proteolysis regulation [1, 2]. This lipocalin is expressed predominantly in the nervous system (NS), both prenatally (vascular pericytes) and postnatally (glia and neurons) and in adulthood [3, 4]. It is also expressed in other tissues and is carried by high-density lipoprotein (HDL) in plasma, so it could interfere in cholesterol and other lipid regulation [5].

ApoD increases considerably in systemic apocrine gland tumors and also in some primary brain tumors [6]. Although the specific biological role of ApoD is unknown, the presence of ApoD in tumors appears to be a prognostic factor in their evolution. Regarding the NS, increased ApoD expression observed in many neurodegenerative diseases could be used to make an early diagnosis thereof [6].

Under cellular stress conditions, ApoD presents extra- and intracellular overexpression [7, 8], suggesting that it plays a fundamental role in cell proliferation, survival, and death.

2. Structure and metabolism of ApoD

ApoD gene is located on human chromosome 3 and chromosome 16 in rodents. Its amino acid sequence does not maintain similarity to other apolipoproteins but is highly similar to some members of the lipocalins family [6, 9].

The molecular weight of mature human ApoD is 19 kDa. It consists of 169 amino acids with glycosylation sites at residues 45 and 178, corresponding to asparagine. Its molecular weight is 32 kDa calculated by sodium dodecyl sulfate-polyacrylamide gel electrophoresis (SDS-PAGE), and approximately 18% of them are carbohydrates. The glycosylation pattern of ApoD varies depending on the site; these values correspond to plasma ApoD, where carbohydrates are less complex and extensive and glycosylation is therefore smaller than ApoD in other body secretions and tissues [6].

Secondary structure has been proposed as a small β-barrel structure constituted by eight antiparallels β-leaves (**Figure 1B**). Within this framework, hydrophobic residues are situated in

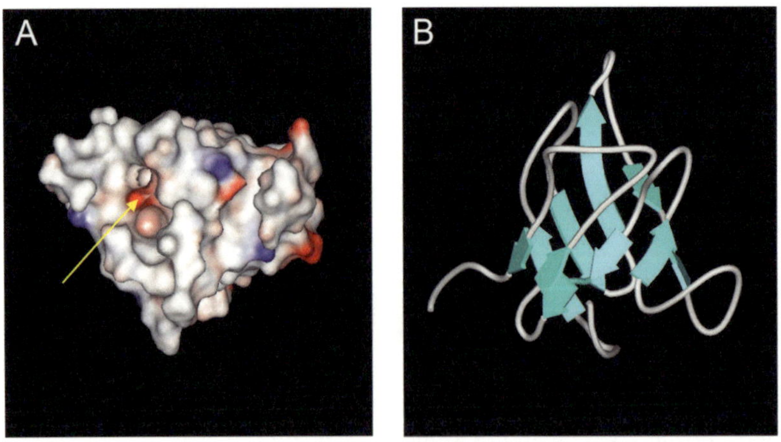

Figure 1. Tertiary structure of apolipoprotein D in three-dimensional vision. **A.** Model of protein surface. The pocket where hydrophobic ligands bind is indicated with an arrow. **B.** Model of α-carbon skeleton of the protein, also showing the eight β-chains that constitute the barrel of this lipocalin. Both figures have been set up with the program Weblab ViewerLite.

http://dx.doi.org/10.5772/66626

	Human	Rabbit	Guinea pig	Rat	Mouse	*E. coli*
Human	–	80%	78%	69%	71%	31%
Mouse	71%	82.6%	–	92.6%	–	–

Table 1. Similarity of the amino acid sequence in apolipoprotein D found in different species.

the inner surface of a central pocket (arrow in **Figure 1A**). This location would be the binding site for the ApoD ligand. Other hydrophobic residues could also participate in the association of ApoD with HDL particles [6].

It has been found both on a molecular as gene scale that ApoD is a heterogeneous protein. It has different isoforms and there is evidence of the presence of two alleles that are expressed in a codominant way in a single gene locus. The first fact can be explained, partly, by a posttranslational process, consisting sometimes in the addition of sialic acid. With respect to the gene, two different alleles were identified by digestion with enzymes Taq1 and Msp1.

Population studies show variations in ApoD gene as polymorphisms that may affect the function of ApoD, the lipoprotein metabolism, and plasma concentrations thereof. In fact, certain ApoD gene alleles show a significant correlation with the predisposition to certain neurodegenerative diseases [10].

The ApoD is also found in various mammals, with very similar functions to human ApoD, and whose similarity in amino acids is shown in **Table 1**. But it is found not only in mammals but also in birds with even greater similarity to human lipocalin than that of some mammals. There are homologous genes in insects and is even located in prokaryotes. In *Escherichia Coli*, there is a lipocalin (Blc) present in the outer membrane of the bacteria that maintain a 31% similarity with human ApoD. It is the first lipocalin located in bacteria and its expression occurs mainly during the stationary phase interacting with the response to hunger during this phase of bacterial cycle.

2.1. Location, synthesis, and expression of ApoD

ApoD has been detected in a variety of organs, tissues, and fluids, reflecting its importance and suggesting that it may play different roles depending on the organ in which it is located. It has been detected in plasma, tear fluid, in the eye ciliary body, in the cerebrospinal fluid (in concentrations without relation to plasma concentration), in the perilymph (in similar concentration to plasma), in the middle ear fluid, in urine, and in sweat [6].

Among the mammals, many interspecific differences in the ApoD expression can be found in different organs and tissues (see **Table 2**).

In contrast to other lipoproteins, the main synthesis of ApoD is not produced in the intestine and liver but in the adrenal glands, kidney, and central nervous system (CNS) [6]. Cells expressing as many ApoD mRNAs are perivascular fibroblasts, glial cells, pial and perivascular CNS, and some neurons [3, 4].

	Human	Monkey	Rabbit	Guinea pig	Rat	Mouse	Hamster
Brain	+	+	+	+	+	+	+
Liver	+	+	nd	+	+	+	−
Kidney	+	+	nd	+	+	-	−
Intestine	+	+	nd	nd	+	−	−
Pancreas	+	+	nd	nd	+	nd	nd
Placenta	+	nd	nd	nd	nd	nd	nd
Adrenal	+	nd	+	+	+	+	nd
Spleen	+	+	+	+	+	−	−
Heart	nd	nd	nd	nd	+	−	+
Bladder	nd	+	nd	nd	+	nd	nd
Skin	nd	+	nd	nd	+	nd	nd
Lung	nd	nd	+	+	+	+	nd
Testicles	nd	+	+	+	+	+	nd
Ovaries	nd	nd	+	+	−	+	nd

+: presence; −: Not present; nd: not determined.

Table 2. Differences in the expression of ApoD in different organs and tissues of several mammals.

2.2. Ligands and functions of ApoD

The ApoD is mainly part (83%) of the high density lipoproteins (HDL) [11], but can also be found in small amount in very low-density (VLDL) and low-density lipoproteins (LDLs) [6]. As a component of HDL, it was observed that ApoD is associated to cholesterol ester transferase protein (CETP), to Apo AI or Apo A-II (over 50% of ApoD present in HDL is forming part of these complexes with Apo A-II) [11]. In HDL, ApoD is also forming part of the complex responsible for the transport of cholesterol from peripheral tissues to the liver for its metabolism, especially with the lecithin-cholesterol acyltransferase (L-CAT). It is believed that ApoD could stabilize the enzymatic activity of the L-CAT or act as a substrate or reaction products carrier, such as cholesterol or cholesteryl esters, as an increase in the activity of cholesterol esterification by L-CAT in the presence of ApoD has been observed. All this suggests that interactions between cholesterol metabolism and ApoD exist, but there is evidence that cholesterol is not the main ligand of ApoD as initially believed. This is supported by the low affinity existing between them and by the fact that in the cyst fluid of breast cancer ApoD concentration increases up to 1000 times, while cholesterol increases only twice [6].

The ApoD can also be free or bound to other small molecules as it interacts with many ligands such as progesterone and other progestins, pregnenolone, bilirubin, arachidonic acid (AA), estrogens, androgens, and E-3-methyl-2-hexanoic acid (major component of underarm odor). Of all these, ApoD is the molecule that has a higher affinity for arachidonic acid, which make us believe that through the L-CAT, ApoD could join in the metabolism regulation, removing it

http://dx.doi.org/10.5772/66626

in order to prevent its transformation into cholesterol esters [6]. It is also important to mention that other linking molecule is retinoic acid, a molecule that plays a major role in the development of nervous system. Indeed, ApoD is the human lipocalin with higher affinity for retinoic acid [12].

Belonging to the family of lipocalin and the variety of tissues in which ApoD is expressed, Makes us pose the hypothesis that this apolipoprotein is multiligand and multifunction and that both function as ligands vary depending on the organ in which it is expressed [6].

2.3. Expression regulation of ApoD

Transcriptional regulation of the expression of ApoD seems to be very complex due to the many factors that modulate this protein [13]. Overall, we could say that there are changes in cell proliferation which modulate the expression of ApoD, or vice versa. Do Carmo et al. have studied in detail the ApoD promoter and determined the genomic region required for the induction of ApoD when cell senesces (when the crop exceeds the confluence). In this genomic region, there is a purine-pyrimidine fragment alternation and octanucleotide SRE (serum-responsive element) that appear to be essential for the induction. Curiously, SRE sites are present in sites related to cholesterol and fatty acids metabolism and mediate the regulation of transcription of these genes depending on sterol genes [13].

3. ApoD as a protective agent against oxidative stress

To conclude this section, we note that in studies conducted in animal models (particularly in the fruit fly *Drosophila melanogaster* and the mouse), in which the expression of ApoD has been genetically modified, ApoD deficit involves behavioral defects and neuronal death by apoptosis. Also, there is less resistance to NS stimuli that induce oxidative stress (OS). Moreover, lipocalin overexpression leads to increased resistance to factors which induce increased oxidative stress [14–16].

In summary, ApoD is a protein with multiple functions depending on the location in which it is expressed. It is a multiligand protein, but it does not mean that it has a specific role, so many studies are still needed to unravel the functional role of ApoD at different levels in which it operates. It has an important role in CNS pathologies, as it behaves as an acute phase protein, rising in neuronal damage. However, we cannot yet say that it acts as a neuroprotective or neurotoxic protein.

We also checked that oxidative stress induces the expression of ApoD in the nervous system [14]. This and the fact that many of the diseases mentioned above occur with increased oxidative stress made us think that this lipocalin plays an important role in controlling this stress when it occurs in pathological conditions.

Evidence from animal models supports this hypothesis. It is specifically carried out with experiments in *Drosophila* fly and mice. In the fly, the Glial Lazarillo protein (Glaz) is the homologous protein to ApoD. In Glaz mutant flies that inactivate its expression, subjected to a stimulus that induces oxidative stress, neuronal death and degeneration are increased. By contrast, in flies with excess of ApoD there is a greater resistance to oxidative stress [16]. In experiments conducted in knockout ApoD mice (in which the gene is inactivated), we have found that these have also a lower resistance to oxidative stress and behavioral changes [17].

4. ApoD in nervous system and its relationship with neurodegenerative pathologies

We can find ApoD both in the central nervous system (CNS) and in the peripheral nervous system (PNS) and is part of the small group of apolipoproteins that are synthesized in the NS close to Apo E, J and C-I [6].

ApoD's role as a conveyor of lipid molecules suggests that it might play an important role in lipid transport during neuronal regeneration [6].

The ApoD accumulates in the peripheral nerve after its injury. Its concentration is much higher there than other apolipoproteins and also has been shown to be synthesized locally and not from the bloodstream, such as Apo A-I and Apo A-IV. In rat sciatic nerve injury, the concentration is increased 500-fold relative to baseline, and mRNA is elevated up to 40 times. Its mission might be to transport cholesterol to remyelination and new membrane formation. It could also carry bilirubin, which is found in damaged nerves, thereby preventing toxic accumulation of the same [6].

In adult animals, ApoD expression is primarily located in the pial and perivascular cells, astrocytes and oligodendrocytes and, inconsistently, in neurons [6].

ApoD has been observed attached to oligodendrocytes in the white matter of the human cerebral cortex. In the gray matter of young individuals, ApoD expression of both glial cells and neurons is limited; this expression increases with normal aging. It seems that ApoD synthesis could be linked to the phenomenon of cellular activation in astrocytes, taking place in astrogliosis. ApoD is constitutively secreted by mouse astrocyte cultures.

In neurons, labeling studies indicate that ApoD is more abundant in some areas than in others. In cerebrum and cerebellum, its labeling is poor and inconsistent, while in the vestibular nuclei, bulbar olive and raphe marking are abundant and constant.

In the nervous system, ApoD could participate in the process of regeneration and remyelination. It has also been proposed as lipids and other substances carrier through the blood-brain barrier. It could also play a significant role in maintaining appropriate levels of cholesterol in compartments not directly exposed to blood. Finally, as a function of this lipocalin the local transport of steroid hormones has also been proposed, which modulate the formation of synaptic connections [6].

4.1. ApoD and nervous injury

The relation between ApoD processes and nerve regeneration has conducted several studies in which injuries are reproduced in the CNS. In all of them, an increase in the ApoD marking has been observed, quite possibly of a local origin as ApoD concentration in the bloodstream

http://dx.doi.org/10.5772/66626

was low. This may be due to the need of increased lipid traffic during the period of reconstruction and restructuration following serious injuries. However, in a study conducted in deficient ApoE mice subjected to hypoxia To cause a stroke, no change was seen in the levels of ApoD with respect to wild strains, where this apolipoprotein would not be involved in cleaning lipid material in the area of the lesion and not be a part of the compensatory mechanism against the absence of ApoE.

4.2. ApoD and nervous pathologies

There are some nerve pathologies in which ApoD is elevated with respect to normal healthy individuals (see **Table 3**).

4.2.1. Niemann-Pick disease type C

In the mouse, this disease is a hereditary disorder of cholesterol homeostasis, which accumulates in unesterified form in the lysosomes. We found progressive dementia and development of neurofibrillary tangles. It was observed that the amount of ApoD in brains of diseased individuals was greater than that in healthy individuals and it was mainly secreted by astrocytes. This suggests ApoD intervention in cholesterol metabolism, acting as a conveyor of lipids released in demyelinating disease processes.

4.2.2. Alzheimer's disease

This disease is a disease of pathology substrate constituent of senile plaques and neurofibrillary tangles. These lesions are clinically reflected in a consistent mental decline in dementia, disorientation, memory loss, and learning capacity.

In these patients, increased ApoD is found in the hippocampus and CSF compared with control subjects. Also, a correlation was found between ApoD and the presence or absence of the ApoE4 allele, so that high concentrations of ApoD were interpreted as a compensatory mechanism against the absence of a particular allele of ApoE in nerve regeneration and maintenance and CNS repair. The ApoD would act in transporting different substances. This theory has been challenged by studies with ApoE-deficient mice in which the expression of ApoD was not altered by the absence of ApoE.

Other studies indicate an ApoD increase in the entorhinal and temporal cortex in elderly subjects. According to some authors, ApoD could participate in the neurochemical cascade associated with chronic CNS neuronal degeneration. Other studies indicate that there is no correlation between ApoE and ApoD and both are involved in neurodegeneration in this disease independently.

Moreover, the analysis of senile plaques and neurofibrillary tangles has resulted in conflicting data, some authors find ApoD presence while in other cases not. This may be due to the employed antibodies. In case of ApoD found in senile plaques, ApoD suggests an important role in fibrillogenesis and deposition of amyloid peptide. The presence of ApoD in neurofibrillary tangles is little or null although we have found a correlation between the number of tangles and the amount of ApoD in patients. It has been interpreted as the injured cortical neurons increase the expression of ApoD before tangles accumulate inside. The concentration of

Pathology/alteration nervous system	Overexpression site	Detected Increase		Ref.
		mRNA	Protein	
Cerebellar ataxia (two mouse models)	- Cerebellum	+		[18]
Unverricht-Lundborg disease (Mouse model: progressive ataxia)	- Cerebellum	+		[19]
Niemann-Pick disease – type C (mouse model: progressive neurodegeneration, ataxia)	- Cerebellum, fraction myelinated - Globus pallidus, thalamus, substantia nigra. White matter in the internal capsule and cerebellum. oligodendrocyte precursors - Brain	+	+ +	[20] [21] [22]
Alzheimer's disease	- Cerebral spinal fluid - Pyramidal neurons with granulovacuolar degeneration - Cortex with neurofibrillary changes - Hippocampus - Deposits β-amyloid	+	+ + + + +	[23] [24] [25] [23, 26] [27]
Schizophrenia and/or bipolar disorder	- Blood serum - Caudate and Brodmann area 9		+ +	[7, 28] [7]
CNS demyelinating diseases (multiple sclerosis)	- Spinal fluid and serum (Intrathecal production)		+	[29]
Astrocytomas (pilocytic and other noninfiltrating)	- Astrocytoma	+	+	[30]
Transmissible spongiform encephalopathies (Mouse)	- Brain	+		[31]
lethal Sindbis virus encephalitis (mouse)	- Central nervous system	+		[32]
Response to neuroleptic drugs (mouse)	- Fluted, globus pallidus, thalamus, and white matter	+	+	[7, 28, 33]
Neuropathic pain after spinal nerve ligation L5 and L6 (rat)	- Dorsal root ganglion	+		[32]
SNP damage (rat): - section sciatic nerve	- site of injury - Fibroblasts in the perineural space	 +	+ 	[34] [35]
CNS damage (rats): - Entorhinal cortex injury - Kainic acid in hippocampal CA layer	- Ipsilateral Hippocampus - Pyramidal neurons in the lesion	 + 	 + +	 [36] [37]
Aging under normal conditions	- reactive astrocytes in cerebral cortex	+	+	[24, 38, 39]

Table 3. Pathological situations or cell damage which overexpresses the ApoD gene and/or protein accumulates in the SN.

this protein in the hippocampus appears to be related to the severity of intraneuronal neurofibrillary changes, but not with extracellular amyloid peptide level. Thus, the most advanced patients (according to the scale of Braak) have a higher content of ApoD.

ApoD has also been located in the vascular preamyloid and amyloid deposits of cerebral amyloid angiopathy, present in most Alzheimer's patients and which is common in the elderly. These patients are at an increased risk of vascular rupture. An inverse behavior has been shown between ApoD and ApoE which would indicate that both proteins have different roles in the development of the disease.

It has also been shown the existence of an increased ApoD expression in rats expressing the mutated protein of human amyloid peptide precursor (characteristic of this disease). These changes were most striking in the hippocampus fimbria, corpus callosum, and other white matter tracts. This may represent a compensatory glial response to amyloid peptide deposition in Alzheimer.

4.2.3. *Spongiform encephalopathies*

An increase in ApoD expression has been shown, especially in later stages, possibly as a result of cellular stress.

4.2.4. *Demyelinating diseases*

In all of them, the elevation of ApoD is found in CSF, possibly due to rupture of the blood-brain barrier, but showed an inconstant behavior in relation to the plasma protein levels. In multiple sclerosis, there is also an increase in ApoD intrathecal production in the early stages of the disease. This has been considered a consequence of demyelination and remyelination processes that characterize the disease in the early stages. Also, increased ApoD has been described in patients treated with steroid.

4.2.5. *Schizophrenia and bipolar disorder*

In these psychiatric disorders, the expression of ApoD also increases in serum and brain, acting as disease marker. In schizophrenia and schizoaffective disorders, a decrease of arachidonic acid is found in the membrane of blood cells, fibroblasts, and brain tissues. We also found calcium-independent phospholipase A2 increased activity in psychiatric illnesses as a decrease of calcium-dependent phospholipase A2 activity [7]. It has shown an increase in the expression of ApoD in the striatum of rodents treated with clozapine and in various regions of white matter. In control animals, ApoD is mainly expressed in astrocytes while in treated animals the increase occurred mainly in neurons. This indicates a contribution of ApoD to antipsychotic mechanisms of this neuroleptic [6]. In addition, arachidonic acid also increases with treatment [7].

There are regional differences in ApoD expression when comparing schizophrenia with bipolar disorder, whereby ApoD could intervene in a natural response to the targeted effects of this neuropathology. In treated patients with schizophrenia, ApoD levels descend compared to normal values. The low concentration of ApoD in serum confirms the association of ApoD with a systemic deficiency in lipid metabolism. Serum ApoD levels of patients with schizophrenia are greater than those of control subjects, which may indicate the onset of the disease. Arachidonic acid (AA) is the precursor of eicosanoid synthesis and prostaglandin metabolism and relates to the formation of the second messenger cAMP. ApoD could link to transport and union of AA, prevent peroxidation and make it accessible for the synthesis of membrane phospholipids, protecting the neuronal membrane functions [7].

Finally, we should note that in studies carried out in animal models (particularly, in the *D. melanogaster* fly and mouse), in which ApoD expression has been genetically modified, the deficit of ApoD involves behavioral defects and neuronal death by apoptosis. Also, there is less resistance to SN stimuli that induce oxidative stress. Moreover, lipocalin overexpression leads to increased resistance to factors which induce increased oxidative stress [9, 15, 16].

In summary, ApoD is a protein with multiple functions depending on the location in which it is expressed. The fact that it is also multiligand does not allow us to point a specific role for this protein; so, many studies are still needed to unravel the functional role of ApoD at different levels in which it operates. What we can say is that it has an important role in CNS pathologies. We cannot yet determine if it acts as a neuroprotective or neurotoxic protein, but we can affirm that it behaves as an acute phase protein, rising in neuronal damage.

4.2.6. ApoD gene polymorphisms

As mentioned earlier, we have determined four gene polymorphisms ApoD in the patients in our study.

Polymorphism rs1467282, c.334 + 718T > C in intron 4 of the gene was determined as genetic variation ApoD in Alzheimer's disease in Finnish population [10]. The theoretical European population frequencies given in **Table 4** are as NCBI database—single nucleotide polymorphism (http://www.ncbi.nlm.nih.gov/projects/SNP).

The polymorphism rs 5952 c.44T> C showed increase risk of sporadic Alzheimer's disease in Chinese population. Rs rs 1568566 haplotype T 5952 C showed lower risk and could be interpreted as a protective factor against Alzheimer's disease [40]. No data frequency in European population was seen.

With respect to Rs1568565 polymorphism c.124-352A > G (intron 2), the -352G allele was associated with a threefold increase risk of Alzheimer's disease early onset (≤65 years) Finnish population [10]. The theoretical European population frequencies given in **Table 5** are as NCBI database—single nucleotide polymorphism (http://www.ncbi.nlm.nih.gov/projects/SNP):

As we see, ApoD could confer protection against damage in the body by oxidative stress, certain polymorphisms ApoD could act in these diseases protecting the carriers of said damage. That is why we think that variations of this gene may confer susceptibility to damage caused by oxidative stress.

Genotype/allele	CC	CT	TT	C	T
Frequency	0.860	0.116	0.023	0.919	0.081

Table 4. Data of theoretical frequencies of rs1467282 in European population.

Genotype/allele	CC	CT	TT	C	T
Frequency	0.833	0.125	0.042	0.896	0.104

Table 5. Data of theoretical frequencies of rs1568565 in European population.

5. Vascular actions of ApoD

As mentioned above, the 1–2% of HDL is formed by ApoD. Likewise, we can say that 83% of plasmatic ApoD is found forming part of HDL particles. Thus, it has been found that decreasing ApoD in HDL before it descends in serum increases the risk of stroke. The fact that the HDLs are responsible for transporting cholesterol from peripheral tissues to the liver and that most ApoDs are present in these lipoproteins makes us think about its beneficial effects from the cardiovascular standpoint [11].

The presence of ApoD has been detected in the atheromatous plaque, but not in normal coronary arteries [11]. An important finding is the large amount of ApoD in quiescent cells [13]. However, its expression is greatly reduced in proliferating cells. In vitro studies show that ApoD inhibits cell proliferation, obtaining a similar effect to that produced by calcium antagonists, demonstrating its beneficial effects in the cardiovascular field. Also in his role of inhibiting proliferation could play an important role in cancer [11].

It has been observed that ApoD is related to the cell migration of vascular smooth muscle for closing the ductus arteriosus and the platelet-derived growth factor—BB (PDGF-BB)—may mediate its expression and localization. It has also been found that ApoD is necessary for the migration of pulmonary artery in response to PDGF-BB.

6. ApoD and tumor pathology

ApoD is associated with reduced proliferative activity of cancer cells, and is abundantly raised in senescent cells. In breast cancer, ApoD expression is associated with favorable histology and clinical stage, whereas in adjacent tumor stromal ApoD expression is a marker of adverse prognosis. Estrogen receptor expression in breast cancer is inversely related to ApoD expression. Therefore, a combined estrogen receptor positivity/ApoD positivity could reflect a nonfunctional estrogen receptor pathway, and this subset of breast cancer patients does not react to adjuvant tamoxifen treatment [41].

Our group has conducted studies in the field of colorectal cancer (CRC) and ApoD. In CRC, tumor growth coincides with increased inflammation, COX-2, and nuclear factor kappa B-mediated, which triggers the release of tumor necrosis factor alpha and interleukin-6 [42]. This inflammation increment corresponds to increased levels of reactive oxygen species (ROS) and their reactive derivatives, inducing in turn oxidative stress (OS) [43–45]. CRC cells show an increase in lipid peroxidation by products that could be triggered by the increased arachidonic acid levels attained by the increased activity of COX-2 [46, 47]. The accumulation of lipid peroxidation results in cell damage and death. However, cancer cells tend to reduce the levels of the ant proliferative cytokine TGF-β1 and the lipid peroxidation adduct 4-hydroxynonenal (4-HNE) as a way to prevent apoptosis.

We have seen that ApoD is related to protection against oxidative stress. It has also been linked to decreased cell proliferation in models [11, 48]. Its expression is regulated by p73 and p63, both p53 family members, which is tumor suppressor so that its expression shows an inverse correlation with tumor growth [49].

As we can see, ApoD levels are increased in the oxidative stress as a defense mechanism but in turn are reduced in advanced stages of cancer. Our group has studied mechanisms that contribute to this paradox and the influence of ApoD in cancer progression and patient survival.

Our results show a repression of ApoD gene expression in CRC, particularly in the initial stages of the disease, which correlates with an elevation of lipid peroxide, adducts in the tissue. In normal mucosa, ApoD protein is present in *lamina propria* and enteroendocrine cells. In CRC, ApoD expression is heterogeneous, with low expression in stromal cells commonly associated with high expression in the dysplastic epithelium. ApoD promoter is basally methylated in HT-29 cells but retains the ability to respond to OS. Exogenous addition of ApoD to HT-29 cells does not modify proliferation or apoptosis levels in control conditions, but it promotes apoptosis upon paraquat-induced oxidative stress [50].

Our results show ApoD as a gene responding to oxidative stress in the tumor microenvironment. Besides using ApoD as marker of initial stages of tumor progression, it can become a therapeutic tool promoting death of proliferating tumor cells suffering oxidative stress [50].

Author details

Jesús Crespo-Sanjuán*, Nuria Zamora-Gonzalez, María Dolores Calvo-Nieves and Cristina Andres-Ledesma

*Address all correspondence to: pmpeius@hotmail.com

Department of Clinical Laboratory, University Clinical Hospital of Valladolid, Valladolid, Spain

References

[1] Flower DR. The lipocalin protein family: structure and function. Biochem J. 1996; 318, 1–14.

[2] Flower DR, North ACT, Sansom CE. The lipocalim protein family: structural and sequence overview. Biochim et Biophys Acta (BBA). Protein Struct Mol Enzymol. 2000; 1482, 9–24

[3] Sánchez D, Ganformina MD, Martinez S. Expression pattern of the lipocalin apolipoprotein D during mouse embryogenesis. Mech Develop. 2002; 110, 225–9.

[4] Ganfornina MD, Sánchez D, Pagano A, Tonachini L, Descalzi-Cancedda F, Martínez S. Molecular characterization and developmental expression pattern of the chicken apolipoprotein D gene: Implications for the evolution of vertebrate lipocalins: Cloning and Characterization of Chicken ApoD. Dev Dyn. 2005 Jan; 232(1), 191–9.

[5] Rassart E, Bedirian A, Do Carmo S, Guinard O, Sirois J, Terrisse L, Milne R. Apolipoprotein D. Biochim Et Biophys Acta. 2000; 1482(1–2), 185–98. Review.

[6] Navarro-Incio AM, Tolivia-Fernandez J. The involvement of apolipoprotein D in pathologies affecting the nervous system. Rev Neurol. 2004; 38(12), 1166–75. Review.

[7] Thomas EA, Dean B, Scarr E, Copolov D, Sutcliffe JG. Differences in neuroanatomical sites of apoD elevation discriminate between schizophrenia and bipolar disorder. Mol Psychiatry. 2003; 8(2), 167–75.

[8] Leung WCY. Apolipoprotein D and platelet-derived growth factor-BB synergism mediates vascular smooth muscle cell migration. Circ Res. 2004 Jun 10; 95(2), 179–86.

[9] Ganfornina MD, Gutierrez G, Bastiani M, Sanchez D. A phylogenetic analysis of the lipocalin protein family. Mol Biol Evol. 2000; 17(1), 114–26.

[10] Helisalmi S, Hiltunen M, Vepsäläinen S, Iivonen S, Corder EH, Lehtovirta M, et al. Genetic variation in apolipoprotein D and Alzheimer's disease. J Neurol. 2004 Aug;251(8):951–7

[11] Sarjeant JM. Apolipoprotein D inhibits platelet-derived growth factor-BB-induced vascular smooth muscle cell proliferated by preventing translocation of phosphorylated extracellular signal regulated kinase 1/2 to the nucleus. Arterioscler Thromb Vasc Biol. 2003 Sep 4; 23(12), 2172–7.

[12] Breustedt DA, Schönfeld DL, Skerra A. Comparative ligand-binding analysis of ten human lipocalins. Biochim Biophys Acta BBA – Proteins Proteomics. 2006 Feb; 1764(2), 161–73.

[13] Do Carmo S, Seguin D, Milne R, Rassart E. Modulation of apolipoprotein D and apolipoprotein E mRNA expression by growth arrest and identification of key elements in the promoter. J Biol Chem. 2002 Feb 15;277(7):5514–23.

[14] Ganfornina MD, Do Carmo S, Lora JM, Torres-Schumann S, Vogel M, Allhorn M, et al. Apolipoprotein D is involved in the mechanisms regulating protection from oxidative stress. Aging Cell. 2008 Aug;7(4):506–15.

[15] Sanchez D, López-Arias B, Torroja L, Canal I, Wang X, Bastiani MJ, et al. Loss of Glial Lazarillo, a homolog of apolipoprotein D, reduces lifespan and stress resistance in Drosophila. Curr Biol. 2006 Apr;16(7):680–6.

[16] Walker DW, Muffat J, Rundel C, Benzer S. Overexpression of a Drosophila homolog of apolipoprotein D leads to increased stress resistance and extended lifespan. Curr Biol. 2006;16(7):674–9.

[17] Ganfornina MD, Do Carmo S, Lora JM, Torres-Schumann S, Vogel M, Allhorn M, et al. Apolipoprotein D is involved in the mechanisms regulating protection from oxidative stress. Aging Cell. 2008 Aug;7(4):506–15.

[18] Diaz E, Ge Y, Yang YH, Loh KC, Serafini TA, Okazaki Y, Hayashizaki Y, Speed TP, Ngai J, Scheiffele P. Molecular analysis of gene expression in the developing pontocerebellar projection system. Neuron. 2002;36(3):417–34.

[19] Lieuallen K, Pennacchio LA, Park M, Myers RM, Lennon GG. Cystatin B-deficient mice have increased expression of apoptosis and glial activation genes. Hum Mol Genet. 2001 Sep 1;10(18):1867–71.

[20] Suresh S, Yan Z, Patel RC, Patel YC, Patel SC. Cellular cholesterol storage in the Niemann-Pick disease type C mouse is associated with increased expression and defective processing of apolipoprotein D. J Neurochem. 1998 Jan;70(1):242–51.

[21] Ong W-Y, Hu C-Y, Patel SC. Apolipoprotein D in the Niemann-Pick type C disease mouse brain: an ultrastructural immunocytochemical analysis. J Neurocytol. 2002 Feb;31(2):121–9.

[22] Yoshida K, Cleaveland ES, Nagle JW, French S, Yaswen L, Ohshima T, et al. Molecular cloning of the mouse apolipoprotein D gene and its upregulated expression in Niemann-Pick disease type C mouse model. DNA Cell Biol. 1996 Oct;15(10):873–82.

[23] Terrisse L, Poirier J, Bertrand P, Merched A, Visvikis S, Siest G, et al. Increased levels of apolipoprotein D in cerebrospinal fluid and hippocampus of Alzheimer's patients. J Neurochem. 1998 Oct;71(4):1643–50.

[24] Kalman J, McConathy W, Araoz C, Kasa P, Lacko AG. Apolipoprotein D in the aging brain and in Alzheimer's dementia. Neurol Res. 2000 Jun;22(4):330–6.

[25] Belloir B, Kövari E, Surini-Demiri M, Savioz A. Altered apolipoprotein D expression in the brain of patients with Alzheimer disease: ApoD in Alzheimer Disease. J Neurosci Res. 2001 Apr 1;64(1):61–9.

[26] Glöckner F, Ohm TG. Hippocampal apolipoprotein D level depends on Braak stage and APOE genotype. Neuroscience. 2003;122(1):103–10.

[27] Navarro A, Del Valle E, Astudillo A, González del Rey C, Tolivia J. Immunohistochemical study of distribution of apolipoproteins E and D in human cerebral beta amyloid deposits. Exp Neurol. 2003 Dec;184(2):697–704.

[28] Mahadik SP, Khan MM, Evans DR, Parikh VV. Elevated plasma level of apolipoprotein D in schizophrenia and its treatment and outcome. Schizophr Res. 2002 Nov 1;58(1):55–62.

[29] Reindl M, Knipping G, Wicher I, Dilitz E, Egg R, Deisenhammer F, et al. Increased intrathecal production of apolipoprotein D in multiple sclerosis. J Neuroimmunol. 2001 Oct 1;119(2):327–32.

[30] Hunter S, Young A, Olson J, Brat DJ, Bowers G, Wilcox JN, et al. Differential expression between pilocytic and anaplastic astrocytomas: identification of apolipoprotein D as a marker for low-grade, non-infiltrating primary CNS neoplasms. J Neuropathol Exp Neurol. 2002 Mar;61(3):275–81.

[31] Dandoy-Dron F, Benboudjema L, Guillo F, Jaegly A, Jasmin C, Dormont D, et al. Enhanced levels of scrapie responsive gene mRNA in BSE-infected mouse brain. Brain Res Mol Brain Res. 2000 Mar 10;76(1):173–9.

[32] Labrada L, Liang XH, Zheng W, Johnston C, Levine B. Age-dependent resistance to lethal alphavirus encephalitis in mice: analysis of gene expression in the central nervous system and identification of a novel interferon-inducible protective gene, mouse ISG12. J Virol. 2002 Nov;76(22):11688–703.

[33] Thomas EA, Danielson PE, Nelson PA, Pribyl TM, Hilbush BS, Hasel KW, et al. Clozapine increases apolipoprotein D expression in rodent brain: towards a mechanism for neuroleptic pharmacotherapy. J Neurochem. 2001 Feb;76(3):789–96.

[34] Boyles JK, Notterpek LM, Anderson LJ. Accumulation of apolipoproteins in the regenerating and remyelinating mammalian peripheral nerve. Identification of apolipoprotein D, apolipoprotein A-IV, apolipoprotein E, and apolipoprotein A-I. J Biol Chem. 1990 Oct 15;265(29):17805–15.

[35] Spreyer P, Schaal H, Kuhn G, Rothe T, Unterbeck A, Olek K, et al. Regeneration-associated high level expression of apolipoprotein D mRNA in endoneurial fibroblasts of peripheral nerve. EMBO J. 1990 Aug;9(8):2479–84.

[36] Terrisse L, Séguin D, Bertrand P, Poirier J, Milne R, Rassart E. Modulation of apolipoprotein D and apolipoprotein E expression in rat hippocampus after entorhinal cortex lesion. Brain Res Mol Brain Res. 1999 Jun 18;70(1):26–35.

[37] Ong WY, He Y, Suresh S, Patel SC. Differential expression of apolipoprotein D and apolipoprotein E in the kainic acid-lesioned rat hippocampus. Neuroscience. 1997 Jul;79(2):359–67.

[38] del Valle E, Navarro A, Astudillo A, Tolivia J. Apolipoprotein D expression in human brain reactive astrocytes. J Histochem Cytochem Off J Histochem Soc. 2003 Oct;51(10):1285–90.

[39] Eddleston M, Mucke L. Molecular profile of reactive astrocytes--implications for their role in neurologic disease. Neuroscience. 1993 May;54(1):15–36.

[40] Chen Y, Jia L, Wei C, Wang F, Lv H, Jia J. Association between polymorphisms in the apolipoprotein D gene and sporadic Alzheimer's disease. Brain Res. 2008 Oct;1233:196–202.

[41] Søiland H, Søreide K, Janssen EAM, Körner H, Baak JPA, Søreide JA. Emerging concepts of apolipoprotein D with possible implications for breast cancer. Cell Oncol Off J Int Soc Cell Oncol. 2007;29(3):195–209.

[42] Kraus S, Arber N. Inflammation and colorectal cancer. Curr Opin Pharmacol. 2009 Aug;9(4):405–10.

[43] Olinski R, Jaruga P, Zastawny TH. Oxidative DNA base modifications as factors in carcinogenesis. Acta Biochim Pol. 1998;45(2):561–72.

[44] Ohshima H, Tatemichi M, Sawa T. Chemical basis of inflammation-induced carcinogenesis. Arch Biochem Biophys. 2003 Sep 1;417(1):3–11.

[45] Itzkowitz SH, Yio X. Inflammation and cancer IV. Colorectal cancer in inflammatory bowel disease: the role of inflammation. Am J Physiol Gastrointest Liver Physiol. 2004 Jul;287(1):G7–17.

[46] Bartsch H, Nair J. Potential role of lipid peroxidation derived DNA damage in human colon carcinogenesis: studies on exocyclic base adducts as stable oxidative stress markers. Cancer Detect Prev. 2002;26(4):308–12.

[47] Schmid K, Nair J, Winde G, Velic I, Bartsch H. Increased levels of promutagenic etheno-DNA adducts in colonic polyps of FAP patients. Int J Cancer. 2000 Jul 1;87(1):1–4.

[48] Do Carmo S, Levros L-C, Rassart E. Modulation of apolipoprotein D expression and translocation under specific stress conditions. Biochim Biophys Acta. 2007 Jun;1773(6):954–69.

[49] Dijk WV, Carmo SD, Rassart E, et al. The Plasma Lipocalins α1-Acid Glycoprotein, Apolipoprotein D, Apolipoprotein M and Complement Protein C8γ. In: Madame Curie Bioscience Database [Internet]. Austin (TX): Landes Bioscience; 2000-2013. Available from: https://www.ncbi.nlm.nih.gov/books/NBK6237/

[50] Bajo-Grañeras R, Crespo-Sanjuan J, García-Centeno RM, Garrote-Adrados JA, Gutierrez G, García-Tejeiro M, et al. Expression and potential role of apolipoprotein D on the death-survival balance of human colorectal cancer cells under oxidative stress conditions. Int J Colorectal Dis. 2013 Jun;28(6):751–66.

Section 2

Genetic Variations of Lipoprotein Metabolism

Chapter 3

Effects of Genetic Variations on Lipoprotein Metabolism in Cardiovascular Diseases

Chesa G. Chauke, Zandisiwe E. Magwebu and Jürgen V. Seier

Additional information is available at the end of the chapter

http://dx.doi.org/10.5772/67138

Abstract

Advances in molecular techniques have shown that genetic factors predispose individuals to cardiovascular diseases (CVD). These techniques have made it possible to identify disease-causing genes, prediction to disease susceptibility and responsiveness to drug interventions. For the purpose of this review, therapeutic intervention (niacin) was conducted in a nonhuman primate model to assess the impact of six coincident single nucleotide polymorphisms (cSNP) identified in prioritised reverse cholesterol transport (RCT) and high-density lipoprotein (HDL) metabolism genes. Gene expression findings confirmed that these genetic variants may have a direct impact on the RCT pathway and drug intervention (niacin) response.

Keywords: cardiovascular disease (CDV), candidate genes, HDL-C metabolism, sequence variants, reverse cholesterol transport (RCT)

1. Introduction

Cardiac and vascular complications are complex multifactorial pathologies and difficult to prevent since are associated with both genetic and environmental factors [1]. Research on cardiovascular diseases (CVDs) is constantly evolving and the current focus is directed towards lipid metabolism, molecular and cellular mechanisms, as well as preventive strategies. Most research in the field of lipid metabolism is motivated by an interest to understand normal lipid transport and preventative measures for atherosclerosis abnormalities [2]. Specific genes and apolipoproteins that are involved in lipid metabolism and lipoprotein synthesis have been isolated, sequenced and mapped in the human genome [3]; however, their role in the lipid metabolism and lipid transport can only be inferred by physiological and genetic studies. To determine their overall function, further exploration of genetic alterations must be investigated.

Since the molecular regulation of lipid metabolism and reverse cholesterol transport (RCT) pathway is complex, numerous studies in humans, animals and *in vitro* have been focusing on the protective action of high-density lipoprotein cholesterol (HDL-C), RCT and cholesterol efflux, which can also be augmented for potential therapeutic strategies of CVDs [4, 5].

2. Overview of RCT and cholesterol efflux

The RCT pathway represents an important process involving the transfer of excess cholesterol by HDL particles to the liver for excretion. The ability of HDL to remove cholesterol from cells such as macrophages is linked to the anti-inflammatory and immunosuppressive functions of this lipoprotein [6]. However, the functionality of HDL is impaired in humans with chronic inflammatory diseases and this causes a reduction in the anti-inflammatory and cholesterol transport properties. Studies have shown that apolipoprotein A-I (ApoA-1), lecithin-cholesterol acyltransferase (LCAT), ATP-binding cassette transporter A1 (ABCA1) and scavenger receptor class B type 1 (SR-B1) serve as important cofactors for a number of RCT pathway constituents [7]. The initial step of the pathway involves ApoA-1 being produced by the liver and released into the plasma where it is involved in all stages, including the formation of nascent HDL particles, HDL remodelling by LCAT and delivery of HDL cholesterol directly to the liver via SR-BI or indirectly via CETP-mediated transfer to apoB-containing lipoproteins [8]. Through this process, cholesterol efflux is promoted from the macrophages via ABCA1 and also by the ABCG1 transporter using the action of LCAT (**Figure 1**).

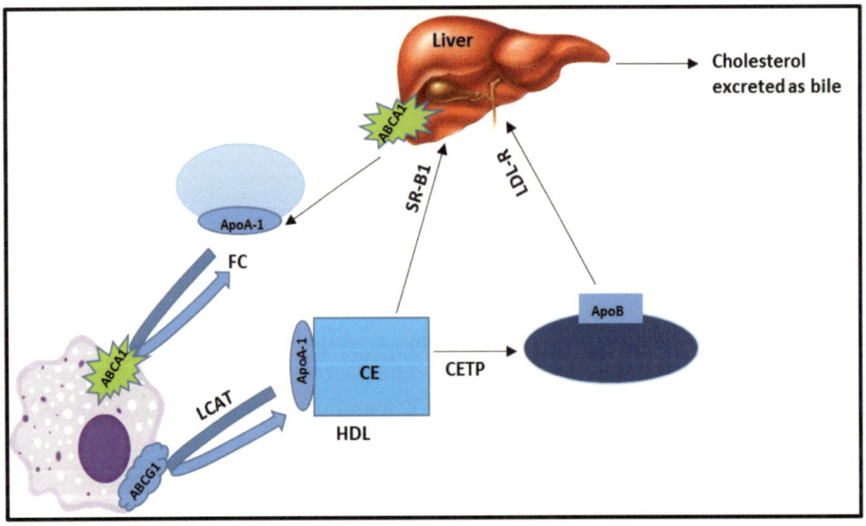

Figure 1. Reverse cholesterol transport (RCT) pathway. Major components of RCT include apolipoprotein A-I (apoA-I), high-density lipoprotein (HDL), lecithin: cholesterol acyltransferase (LCAT), ATP-binding cassette transporter (ABCA1/ABCG1) and cholesterol ester transfer protein (CETP). Free cholesterols (FC) in the HDL are delivered to the liver for excretion through scavenger receptor B1 (SR-B1). Alternatively, the cholesteryl esters (CE) could also be delivered to the liver through the low-density-lipoprotein receptor (LDLR). Figure modified from Rader et al. [4].

http://dx.doi.org/10.5772/67138

3. Vervet monkey as an animal model for CVDs

As with most areas of human biology, studies of human CVDs have been enriched and complemented by investigations of animal models. Among the nonhuman primates (NHP), the vervet monkey (*Chlorocebus aethiops*) has been validated to be an excellent research model for the study of CVDs [9]. For the purpose of this investigation, we used this NHP model to determine the protective action of HDL, its role in the reverse cholesterol transport (RCT) pathway and the expression profile of genes regulating HDL metabolism.

This study was conducted in compliance with the Public Health Service (PHS) Policy on Humane Care and Use of Laboratory Animals (A5726-01) and approved by the Ethics Committee of the South African Medical Research Council (*SA*MRC) (REF 11/07). The selected subjects (25) were healthy adult female monkeys with normal plasma HDL. All individuals were kept under identical housing conditions according to the South African National Standard for the Care and Use of Animals for Scientific Purposes (The SANS 10386:2008).

3.1. Laboratory analysis

3.1.1. Candidate genes and sequence variants selection

The genetic variations were evaluated in 10 genes implicated in lipid metabolism (CETP, ABCA1, CYP7A1, apoA-1, apoB, apoE, SR-B1, LCAT, apoCI and apoCII). Twenty-two coincident single nucleotide polymorphisms (cSNPs) were selected for genotyping. These cSNPs were prioritised based on their function and location within their respective candidate gene and their association with CVD.

3.1.2. Gene expression

Blood (2 ml) was collected in EDTA-containing tubes from 25 animals using a femoral venepuncture after ketamine anaesthesia at 10 mg/kg bodyweight. DNA was extracted from whole blood using the Nucleospin Genomic Blood DNA Purification Kit (MACHEREY NAGEL, Germany) and PAXgene Blood RNA Kit (PreAnalytiX, Qiagen) was used for RNA extraction. The extracted DNA was used for Sanger sequencing while RNA was for gene expression experiments. Turbo DNase treatment (Ambion, USA) was used for RNA purification before cDNA conversion (high-capacity cDNA kit, Applied Biosystems, USA). The effects of niacin treatment on the expression of the 10 prioritised genes were determined using quantitative real-time PCR (qRT-PCR). The gene expression data were normalised to the average of phosphoglycerate kinase 2 (PGK2: QT00219023) and glyceraldehyde-3-phosphate dehydrogenase (GAPDH: QT01192646), which were used as housekeeping genes.

Since the levels of high-density lipoprotein-cholesterol (HDL-C) are a significant determinant of a cholesterol efflux capacity, a correlation analysis was conducted to determine the relationship between the levels of HDL-C and mRNA expression of the 10 selected RCT candidate genes.

4. Effects of mutations, drug intervention and gene expression on RCT

A major area in HDL-based therapeutics is focusing on the development of pharmacological approaches to improve the activity of the RCT pathway. One of these strategies involves the combination of genetic variations and individual responsiveness to drugs [10]. Subsequently, genetic variations in gene encoding transporters contribute to individual differences in drug absorption, elimination and cellular uptake, thereby affecting drug response and toxicity [11].

Among several types of genetic variations, single-nucleotide polymorphisms (SNPs) are the most abundant throughout the genome [12]. SNPs have lately received much attention as they serve as markers of individual risk for adverse drug reactions or susceptibility to complex diseases [13]. Small-scale studies have focused on the effects of polymorphisms on physiological or biochemical factors and have provided useful information on possible mechanistic links between variation at the gene level and risk factors for CVDs [14]. For the purpose of this review, 10 'candidate' genes known to be involved in RCT and HDL metabolism were screened in the vervet model and only six cSNPs (I405V, I883M, A233S, cL96R, -62A>C and A350A) were identified in CETP, ABCA1, CYP7A1, apoCII and SR-B1, respectively (**Table 1**).

Gene	cSNP	Accession number	Chr	Exon	Nucleotide change	Amino acid change
CETP	I405V	rs5882	16	14	A/G	I/V
ABCA1	Ile883Met	rs4149313	9	18	A/G	I/M
CYP7A1	Asn233Ser	rs8192874	8	3	A/G	N/S
APOC-II	Leu96Arg	rs5167	19	3	T/G	L/R
	-62A>C	rs2288911		Promoter		
SR-B1	A350A	rs5888	12	8	C/T	A/A

Table 1. cSNPs identified in vervet monkeys.

For effective changes in lipid metabolism, niacin, as the most potent available lipid-regulating drug [7] was used as a tool to increase HDL levels (**Figure 2**). A strong inverse correlation was observed with CETP, SR-B1 and CYP7A1 concentrations (r = -0.14, -0.27, -0.30; $p < 0.001$). Concurrently, gene expression profile showed that all three genes were down-regulated when correlated with the three cSNPs (I405V, A350A and A233S) (**Figure 3**). Since I405V is known to lower plasma CETP concentration and elevate HDL-C concentration [5, 9], the presence of this variant confirmed the same effect in the vervet model. With a similar expression profile observed in SR-B1 and CYP7A1, the presence of A350A may suggest a possible influence on RCT and HDL-C synthesis and a plausible involvement of A233S in drug metabolism [6]. The remaining cSNPs (I883M, cL96R and -62A>C), however, did not influence the expression of their respective genes (ABCA1 and APOCII) despite being known to alter plasma lipid levels and influence cholesterol efflux [15, 16]. Therefore, these findings suggest that some of these identified sequence variants have significant impact on gene expression which can be correlated with biochemistry levels (HDL-C, LDL-C and triglycerides) following drug intervention.

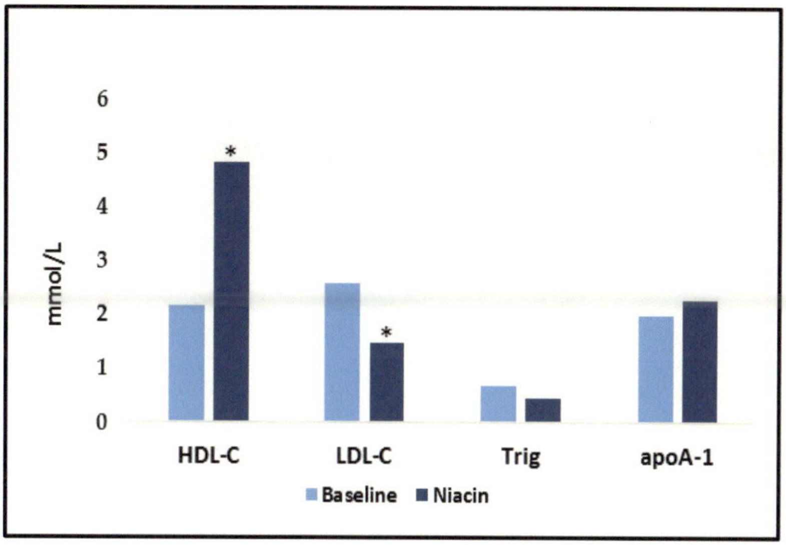

Figure 2. The effect of niacin treatment on HDL-C, LDL-C, triglycerides (Trig) and apoA-1. *Significant level ($p < 0.05$).

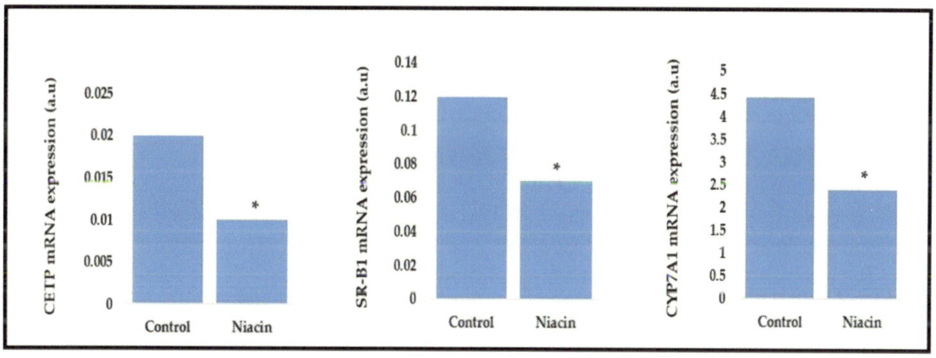

Figure 3. The effect of niacin treatment on CETP, SR-B1 and CYP7A1 mRNA expression. The data were expressed as mean ± SD and mRNA expression in a.u. (arbitrary units). *Significant level ($p < 0.05$).

5. Conclusion

It is a fact that characterisation of polymorphisms in lipid metabolism is challenging, however it remains essential for the optimal regulation and functioning of the RCT pathway. This review demonstrates that the genetic determinants of lipid transport and metabolism may provide additional significant benefit in pharmacological therapy for CVDs. Genetic approaches have shown that sequence variants can be correlated with biochemistry levels such as HDL-C, LDL-C and triglycerides following drug intervention. Although cholesterol lowering alone may explain the anti-atherosclerotic effect of niacin on HDL-C, in this review, gene expression data has shed some light in supporting the hypothesis that genetic variants may influence the expression of genes involved in RCT, which may also play a role in the anti-atherosclerotic effect of niacin.

It is also noteworthy that this is the first report to provide data of a controlled pharmacological intervention linked to genetic determinants of lipid metabolism in vervet monkeys.

Author details

Chesa G. Chauke*, Zandisiwe E. Magwebu and Jürgen V. Seier

*Address all correspondence to: cchauke@mrc.ac.za

South African Medical Research Council, Primate Unit and Delft Animal Centre, Cape Town, South Africa

References

[1] Zaragoza, C., C. Gomez-Guerrero, J.L. Martin-Ventura, L. Blanco-Colio, B. Lavin, B. Mallavia, C. Tarin, S. Mas, A. Ortiz and J. Egido. Animal models of cardiovascular diseases. *BioMed Research International,* 2011. p. 1–13.

[2] Cruz, D.E., H.M. Ahmed, S.R. Jones, M.B. Elshazly and S.S. Martin. Discordance in lipid measurements: Can we capitalize to better personalize cardiovascular risk assessment and treatment? *Current Cardiovascular Risk Reports,* 2014. **8**(5): p. 1–8.

[3] Breslow, J. Lipoprotein transport gene abnormalities underlying coronary heart disease susceptibility. *Annual Review of Medicine,* 1991. **42**(1): p. 357–371.

[4] Rader, D.J., E.T. Alexander, G.L. Weibel, J. Billheimer and G.H. Rothblat. The role of reverse cholesterol transport in animals and humans and relationship to atherosclerosis. *Journal of Lipid Research,* 2009. **50**(Supplement): p. S189–S194.

[5] Assmann, G. and A.M. Gotto, Jr. HDL cholesterol and protective factors in atherosclerosis. *Circulation,* 2004. **109**(23 Suppl 1): p. III8–14.

[6] Yvan-Charvet, L., N. Wang and A.R. Tall. Role of HDL, ABCA1 and ABCG1 transporters in cholesterol efflux and immune responses. *Arteriosclerosis, Thrombosis and Vascular Biology,* 2010. **30**(2): p. 139–143.

[7] Daniels, T.F., X.-L. Wu, Z. Pan, J.J. Michal, R.W. Wright Jr., K.M. Killinger, M.D. MacNeil and Z. Jiang. The reverse cholesterol transport pathway improves understanding of genetic networks for fat deposition and muscle growth in beef cattle. *PloS One,* 2010. **5**(12): p. e15203.

[8] Rothblat, G.H. and M.C. Phillips. High-density lipoprotein heterogeneity and function in reverse cholesterol transport. *Current Opinion in Lipidology,* 2010. **21**(3): p. 229.

[9] Moghadasian, M.H., J.J. Frohlich and B.M. McManus. Advances in experimental dyslipidemia and atherosclerosis. *Laboratory Investigation,* 2001. **81**(9): p. 1173–1183.

[10] Vergani, C., T. Lucchi, M. Caloni, I. Ceconi, C. Calabresi, S. Scurati and B. Arosio. I405V polymorphism of the cholesteryl ester transfer protein (CETP) gene in young and very old people. *Archives of Gerontology and Geriatrics,* 2006. **43**(2): p. 213–221.

[11] Iida, R., T. Yasuda, E. Tsubota, H. Takatsuka, M. Masuyama, T. Matsuki and K. Kishi. M-LP, Mpv17-like protein, has a peroxisomal membrane targeting signal comprising a transmembrane domain and a positively charged loop and up-regulates expression of the manganese superoxide dismutase gene. *Journal of Biological Chemistry,* 2003. **278**(8): p. 6301–6306.

[12] Fareed, M. and M. Afzal. Single nucleotide polymorphism in genome-wide association of human population: a tool for broad spectrum service. *Egyptian Journal of Medical Human Genetics,* 2013. **14**(2): p. 123–134.

[13] Ganji, S.H., V.S. Kamanna and M.L. Kashyap. Niacin and cholesterol: role in cardiovascular disease (review). *The Journal of Nutritional Biochemistry,* 2003. **14**(6): p. 298–305.

[14] Freeman, D.J., N.J. Samani, V. Wilson, A.D. McMahon, P.S. Braund, S. Cheng, M.J. Caslake, C.J. Packard and D. Gaffney. A polymorphism of the cholesteryl ester transfer protein gene predicts cardiovascular events in non-smokers in the West of Scotland Coronary Prevention Study. *European Heart Journal,* 2003. **24**(20): p. 1833–1842.

[15] Chen, C.-H., Y.-L. Cao and W.-C. Hu. Apolipoprotein C-II promoter T→ A substitution at position-190 affects on the transcription of the gene and its relationship to hyperlipemia. *Biochemical and Biophysical Research Communications,* 2007. **354**(1): p. 62–65.

[16] Kyriakou, T., C. Hodgkinson, D.E. Pontefract, S. Iyengar, W.M. Howell, Y.-k. Wong, P. Eriksson and S. Ye. Genotypic effect of the -565C> T polymorphism in the ABCA1 gene promoter on ABCA1 expression and severity of atherosclerosis. *Arteriosclerosis, Thrombosis and Vascular Biology,* 2005. **25**(2): p. 418–423.

Section 3

Inflamatory Conditions and Lipoproteins

Chapter 4

Role of HDL-Associated Proteins and Lipids in the Regulation of Inflammation

Roger White, Samantha Giordano and Geeta Datta

Additional information is available at the end of the chapter

http://dx.doi.org/10.5772/67141

Abstract

Lipoproteins are complexes of lipids and proteins that carry water-insoluble cholesterol in the bloodstream. While cholesterol is required for normal cell function, hypercholesterolemia contributes to the development of cardiovascular disease (CVD). Increased low-density lipoprotein (LDL) is a major risk factor for CVD. Reduced high-density lipoprotein (HDL) levels are inversely related to CVD risk, suggesting a protective role for HDL. Several diseases, including atherosclerosis, diabetes, chronic kidney disease and rheumatoid arthritis, have been identified where HDL levels are decreased or function is compromised. HDLs are spherical particles with a hydrophobic core of cholesteryl esters surrounded by a monolayer of phospholipids, proteins and unesterified cholesterol. Apolipoprotein (apo) A-I, the major protein component of HDL, plays an important role in the assembly and function of HDL. One of the major functions of HDL is to mediate cellular cholesterol efflux and the transfer of cholesterol from extrahepatic tissues to the liver for excretion into the bile. In addition to regulating cholesterol metabolism, HDL also exhibits antioxidative, antithrombotic and anti-inflammatory properties. Under certain conditions, however, HDL may undergo biochemical modification resulting in the formation of a particle with pro inflammatory properties. This review will focus on the variable properties of HDL under normal physiological conditions and in the context of inflammation.

Keywords: HDL, inflammation, lipid composition, protein composition, function, macrophage mitochondria

1. Introduction

Hypercholesterolemia is an important determinant of cardiovascular disease (CVD), the leading cause of death globally [1]. Cholesterol, among other lipids, is carried in the bloodstream from the liver to different parts of the body by lipoproteins, complex particles composed of lipids and proteins. There are four major lipoproteins that can be classified on the basis of their density: chylomicrons, very low-density lipoprotein (VLDL), low-density lipoprotein (LDL) and high-density lipoprotein (HDL) [2]. Chylomicrons, VLDL and LDL are larger particles with densities ranging from 0.95 to 1.063 g/ml. HDL is a mixture of spherical particles ranging in size from 7 to 12 nm in diameter and 1.063–1.21 g/ml in density. Epidemiological studies have established an inverse relationship between HDL cholesterol and CVD risk [3, 4]. Thus, a reduction in plasma HDL levels represents an important risk factor for CVD. Results of clinical trials demonstrate that lowering LDL levels reduces CVD risk [5, 6]. Evidence supporting a role for elevated HDL in reducing CVD risk, however, is still forthcoming. Clinical trials have shown that torcetrapib, dalcetrapib and extended-release niacin significantly increase circulating HDL levels; however, this was not associated with improved outcomes [7–9]. On the other hand, raising plasma HDL by infusion or overexpression of apoA-I in murine models was shown to reduce atherogenic lesion progression [10]. One hypothesis to explain this disparity proposes that the "quality" or functional status of HDL may be a better indicator of CVD risk than plasma levels of HDL per se [11]. This review will focus on the structure-function relationship of HDL and how it influences responses to the lipoprotein in the context of inflammation.

HDL particles have a neutral core of cholesteryl ester and triglycerides (TG) surrounded by a monolayer of phospholipids, free cholesterol (FC) and protein. ApoA-I is the major protein associated with HDL particles and is synthesized in the liver and small intestine. Phospholipids and cholesterol are transferred to apoA-I by a process mediated by the ATP-binding cassette transporter type 1 (ABCA1) [12, 13] resulting in the formation of a lipid poor, dense particle called preβ-HDL. This particle plays an important role in reverse cholesterol transport, a process by which cholesterol is removed from cells. Although these particles have been predominantly studied under in vitro conditions, little information is available regarding the presence or functional significance of preβ-HDL in vivo [14]. HDL isolated from plasma by sequential ultracentrifugation yields two major subpopulations: HDL2, a large, light, lipid-rich particle (d1.063–1.125 g/ml), and HDL3, a smaller, denser protein-rich particle (d1.125–1.21 g/ml). These two particles can be further subdivided into five distinct populations: HDL2b, HDL2a, HDL3a, HDL3b and HDL3c [15]. These heterogeneous particles vary in their lipid and protein composition, forming particles of varying density, charge, and antigenicity. They also possess discrete functional properties.

2. HDL structural components

The HDL lipidome: Phospholipids (PL) represent the major lipid component of HDL, constituting about 50% by weight of all the lipids [15]. Phosphatidylcholine (PC), with a carbon backbone of varying length and saturation, is the major PL species. Lysophosphatidylcholine (LPC), phosphatidylethanolamine (PE), phosphatidylinositol (PI) and plasmalogens are also present at lower, but significant, amounts (greater than 1% of total HDL lipids by weight). Other phospholipids (phosphatidylglycerol (PG), phosphatidylserine (PS), phosphatidic acid (PA) and cardiolipin) constitute less than 1% of total HDL lipids by weight.

Sphingolipids are also well-represented in HDL particles. Sphingomyelin (SM) accounts for 5–10% by weight of total HDL lipids [15]. SM is converted to ceramide by sphingomyelinase [16]. Ceramide constitutes 0.05% by weight of total HDL lipids. Ceraminidase converts ceramide to sphingosine. Finally, the enzyme sphingosine kinase converts sphingosine to sphingosine 1-phosphate (S1P) [16]. S1P, as well as ceramide-1-phosphate, are carried by HDL and are potent signaling molecules that regulate cell growth, survival and differentiation [17]. S1P plays an important role in the suppression of inflammation [17]. S1P binding to HDL requires its physical interaction with apo M [17, 18]. Sphingosylphosphorylcholine and lysosulfatide are additional, biologically active lysosphingolipids carried by HDL [15]. The principal lipids associated with HDL particles are summarized in **Table 1**.

Proteins	**Lipids**
Apolipoproteins (AI-II, A-V, C-I-IV, D, E, F, M, H, O)	Phospholipids:
CETP	PC, PE, PI, PG, PS, PA
PAF-AH	
PLTP	Sphingolipids:
LCAT	SM
PON1, PON3	Ceramides
SAA1, SAA2, SAA4	S1P
Albumin	Sphingosylphosphorylcholine
Transthyretin	Lysosulfatide
Hemoglobin	
Hemopexin	
Transferrin	
Ceruloplasmin	
Vitamin D binding protein	
Complement	

Table 1. Normal protein and lipid components of HDL.

The HDL proteome: The HDL proteome has been characterized by several groups over the past 10 years. Using mass spectroscopy, the presence of at least 85 proteins on HDL have been reported [19]. These fall into different regulatory categories: lipid metabolism, acute phase response (APR), hemostasis, immune response, metal binding, vitamin transport, proteinase inhibitor and complement regulation [19, 20]. A representative list of HDL-associated proteins is shown in **Table 1**. Among these, the lipid metabolism group is the largest and contains apoA-I as well as other apolipoproteins (**Table 1**). As mentioned above, HDL exists as multiple sub-species. The proteins, lecithin-cholesterol acyltransferase (LCAT), phospholipid transfer protein (PLTP) and cholesteryl ester transfer protein (CETP), play a major role in converting HDL from one sub-species to another. APR proteins such as apo A-IV, SAA1 and SAA2 regulate lipid metabolism and are also present along with Apo J, a protein involved in lipid metabolism and complement regulation. Surprisingly, a variety of other proteins with diverse functions such as hemoglobin, hemopexin and transferrin (iron metabolism), ceruloplasmin (metal binding), and vitamin D binding protein (vitamin binding) are also seen. These are described in detail in the review by Shah et al. [19]. Thus, the protein and lipid cargo on HDL significantly influence particle function.

3. Functions of HDL

Reverse cholesterol transport: Under hypercholesterolemic conditions, the accumulation of cholesterol in macrophages leads to the formation of "foam cells" which contribute to atheroma formation. HDL is commonly referred to as the "good cholesterol". The salutary effect of HDL has been attributed to its ability to transfer cholesterol from extra-hepatic tissues to the liver for metabolism and excretion into the bile, a process called reverse cholesterol transport [21]. This is believed to be a critical antiatherogenic function of HDL. Cholesterol from macrophages is transferred to lipid-poor apoA-I [22] *via* ABCA1. The cholesterol is converted to cholesterol esters by the action of LCAT present on HDL. Sequestration of cholesterol esters in the hydrophobic core of the particle is associated with the formation of spherical HDL2 and HDL3. These mature HDL particles also incorporate cholesterol *via* an alternate transporter, the ATP-binding cassette transporter G1 (ABCG1) as well as the scavenger-receptor class B, type 1 (SR-BI) pathway [23]. Cholesterol-enriched HDL is subsequently removed from the circulation by hepatocytes and is excreted by the biliary pathway into bile and feces. In addition to mediating reverse cholesterol transport, HDL also possesses antioxidant, anti-inflammatory and antithrombotic properties. These pleiotropic effects of HDL play a major role in limiting inflammatory injury associated with leukocyte infiltration in the blood vessel wall.

Antioxidant properties of HDL: Chylomicrons, VLDL and LDL are apoB-containing lipoproteins which deliver cholesterol and TG to cells and are strongly implicated in atheroma formation. The response-to-retention hypothesis postulates that [24] LDL is oxidized in the arterial wall by enzymes including myeloperoxidase (MPO), NADPH oxidase, nitric oxide synthase and lipoxygenase, resulting in the accumulation of lipid hydroperoxides (LOOH) [25]. Oxidized LDL (ox-LDL) is taken up by macrophages leading to the formation of foam cells and fatty plaques. Protein and lipid components of HDL inhibit the accumulation of LOOH in LDL and prevent the formation of ox-LDL. LOOH and phosphatidyl choline hydroperoxides (PLOOH) are transferred from LDL to HDL. This process is regulated by the lipid composition and rigidity of the HDL surface. Specifically, HDL surface rigidity is determined by the ratios of SM:PC, FC:PL and saturated to polyunsaturated fatty acids (SFA:PUFA) [26]. Zerrad-Saadi and colleagues have identified the HDL3 particle as a key mediator of LOOH transfer due its optimal surface rigidity and particle content [27].

ApoA-I is likely the major HDL protein species involved in the removal of LOOH moieties from LDL. The methionine (Met) residues 112 and 148 of apoA-I can reduce LOOHs to inactive lipid hydroxides (LOH) [28]. In addition, apoA-I removes seeding LOOH molecules from LDL [29]. In addition to apoA-I, other apolipoprotein and enzyme components of HDL, such as, apo E, apo J, apo A-II, apo L-1, apo F, apo A-IV, PON1/3, PLTP and PAF-AH, play a role in its antioxidant function. Proteomic analyses from the Davidson laboratory [30] demonstrate that HDL3c contains all these proteins along with apo M, apo D, apo A-II, SAA1,2 and 4 and apo C-I and apo C-II. This corroborates earlier studies showing that HDL3c has more potent antioxidant activity than other HDL subspecies [31, 32]. Thus, both lipid and protein components of HDL3c contribute to its antioxidant activity. Kontush et al. [32] have hypothesized that the protein components of HDL3c form a pocket which enables the transfer of LOOH from LDL which is further reduced by the concerted action of apolipoproteins and enzymes in this pocket [26].

***Anti-inflammatory properties of HDL*:** The role of inflammation in atherogenesis has been clearly established [33–35]. Acute and chronic inflammations are associated with monocyte adhesion/infiltration and endothelial cell activation [33–35]. HDL is known to suppress the lipopolysaccharide (LPS)-induced secretion of interleukin-6 (IL-6), tumor necrosis factor-α (TNF-α), interferon-γ (IFN-γ) and other pro-inflammatory mediators [36–38]. HDL also reduces inflammation by neutralizing endotoxin, further supporting its anti-inflammatory role [39]. Thus, HDL exerts its anti-inflammatory effect in multiple ways.

Regulation of endotoxicity: In the context of infection, Gram-negative bacteria release LPS in the circulation which binds CD14 located in membrane rafts on cell surfaces. CD14 engagement facilitates the activation of toll-like receptor 4 (TLR4) binding, resulting in the release of pro-inflammatory cytokines such as IL-6 and TNF-α. HDL is able to inhibit this initial activation step *via* binding to lipid A, a glycolipid component of LPS, thus preventing TLR4 activation. Gram-positive bacteria release lipoteichoic acid (LTA) which, similar to LPS, binds CD14 and activates pro-inflammatory signaling *via* the TLR2/6 pathway [40, 41]. HDL additionally contributes to the inactivation of LPS and LTA by disrupting membrane rafts. In this manner, HDL mediates cholesterol and phospholipid efflux which destabilizes rafts and prevents the assembly of receptor complexes for LPS and LTA [14, 40].

Regulation of macrophage function: Macrophages are a versatile group of cells that play a critical role in regulating immunity, inflammation and lipid metabolism. Macrophage phenotype and function are regulated, in large part, by their environmental milieu [42–45]. On the basis of cell morphology and function, two populations of activated macrophages have been identified [46]. The classically activated M1 macrophage is induced by LPS and Th1 cytokines such as IFN-γ, interleukin-2 (IL-2) and TNF-α [43, 44]. These cells are pro-inflammatory and secrete inflammatory mediators (TNF-α, IL-1, IL-6, IL-15, IL-18, IL-23, IFN-γ), stimulate inducible nitric oxide synthase (iNOS) and promote the formation of reactive oxygen and nitrogen species [47]. The second macrophage phenotype, the alternatively activated M2 macrophage, is induced by IL-4, IL-10, IL-13 and glucocorticoid hormones [42–45]. M2 macrophages play an important role in the resolution of inflammation by inhibiting inflammatory cytokine expression and promoting wound healing [42–45]. HDL and apo A-I have been shown to promote the formation of anti-inflammatory M2 macrophages in human monocyte-derived macrophages [48] and mice [49]. As mentioned in the previous section, HDL3 is a key mediator of reverse cholesterol transport and possesses potent antioxidant properties. Reports from several laboratories suggest that HDL-associated S1P inhibits inflammation *via* activation of the PI3-kinase/*Akt* signaling pathway [50–52]. Pretreatment of bone marrow-derived macrophages (BMDMs) with S1P suppressed LPS-induced secretion of TNF-α, monocyte chemoattractant protein (MCP) and IL-12 [53]. Additionally, Hughes and colleagues reported that S1P enhanced the activity of Arg1 and suppressed the NF-kB-mediated induction of iNOS [53]. These responses to S1P are associated with M2 macrophage polarization.

Regulation of mitochondrial function: The mitochondrion is a double-membraned, energy-producing organelle, which contains its own maternally inherited mitochondrial DNA [54–56]. Under normal conditions, the mitochondrial respiratory chain shuttles electrons through the respiratory complexes, consumes oxygen at Complex IV and pumps hydrogen ions from inside the mitochondria to the intermembrane space at Complexes I, III and IV. This allows ATP production to proceed at the level of Complex V (ATP synthase). Under normal conditions, oxidative phosphorylation is a tightly regulated process with heat and reactive oxygen species (ROS) being produced as byproducts.

In the presence of ox-LDL and other oxidized lipids, the mitochondrion increases the formation of ROS, which can damage the mitochondria and other organelles causing cellular dysfunction and death. HDL, by virtue of its antioxidant properties, can decrease the cellular damage caused by oxidized lipids. The HDL protein PON1 hydrolyzes cholesterol esters and phospholipids in oxidized lipoproteins [52, 57, 58] thus inhibiting mitochondrial damage in the presence of oxidized lipids [58]. Further, HDL-associated apoA-I has been implicated in electron transport chain maintenance and repair [59]. In apoA-I null mice (apoA-I$^{-/-}$), an increase in coronary ischemia-reperfusion injury is observed compared to wild-type mice [59] and is associated with a decrease in the content of the mitochondrial protein Coenzyme Q (CoQ) in cardiomyocytes. CoQ normally supports oxidative phosphorylation by shuttling electrons from Complex II to Complex III. Exogenous administration of CoQ to apo-A-I$^{-/-}$ mice attenuated myocardial infarct size compared to the injury response in untreated mice. These data indicate the importance of HDL, and specifically, apoA-I in preserving mitochondrial structure and function.

Potential mechanisms by which HDL preserves mitochondrial function include activation of the Reperfusion Injury Salvage Kinase (RISK) pathway and the Survivor Activating Factor Enhancement (SAFE) cascade. These are cell survival pathways which are known to prevent mitochondrial damage in models of ischemic pre- and postconditioning [60]. Activation of STAT3 is an important component of the SAFE pathway and results in the downregulation of pro-apoptotic factors Bax and Bad and upregulation of antiapoptotic factor Bcl-2 and the antioxidants manganese superoxide dismutase and metallothionein [60, 61]. Further, STAT3 is transported to the mitochondrion by the GRIM-19 chaperone where it inhibits the release of cytochrome c and reduces cell death [62–64]. In a rodent model of coronary artery occlusion, the administration of apoA-I was shown to decrease infarct size and inhibit mitochondrial morphological changes seen in the heart [60]. Further analyses showed that apoA-I increased the phosphorylation of *Akt* and glycogen synthase kinase 3 beta (GSK3β), known mediators of the RISK and SAFE survival pathways.

The S1P component of HDL is also able to activate the RISK And SAFE pathways [51, 52, 65]. Interestingly, studies conducted in neonatal rat cardiomyocytes showed that S1P is critically required for the phosphorylation of STAT3. In contrast, STAT3 phosphorylation was absent in cells treated with HDL that was deficient in S1P [65]. In addition, S1P stimulates the phosphorylation of the transcription factor, forkhead box O-1 (FOXO-1), which inhibits ROS formation and apoptosis in the phosphorylated form [66, 67]. These data suggest that HDL activates RISK and SAFE pathways and inhibits ROS, mitochondrial dysfunction and cell death.

Interestingly, S1P has also been shown to regulate mitochondrial Complex IV assembly and cellular respiration by interacting with mitochondrial prohibitin-2 (PBH-2) [68]. PBH-2 acts as a scaffolding protein for mitochondria and its interaction with S1P during ischemic preconditioning of cardiomyocytes is essential for cardioprotection [68–70]. These data suggest that S1P can stabilize mitochondrial complexes and inhibit ROS formation, suggesting an alternate cardioprotective mechanism of S1P action.

Recent studies have suggested that other HDL-associated apolipoproteins play a role in preserving mitochondrial structure and function. ApoJ is expressed ubiquitously and is present on small dense HDL3 particles [71–73]. It is considered to be an antioxidant due to the presence of disulfide bonds that inhibit ROS-induced injury and preserve mitochondrial function [74]. Further, apoJ has been implicated in activating *Akt* and GSK3β and the RISK survival pathway [71]. ApoM is found in association with approximately 5% of HDL particles where it confers several cytoprotective properties that include stimulating preβ-HDL formation, facilitating reverse cholesterol transport and inhibiting LDL oxidation [75–78]. ApoM also plays an important role in the cytoprotective response to S1P by binding the sphingolipid and facilitating its incorporation into HDL particles [75, 79, 80]. It follows that overexpression of apoM in mice reduces infarct size in response to ischemia-reperfusion injury and preserves mitochondrial function by increasing the HDL content of S1P.

4. Inflammation-induced alterations in HDL structure

Changes in HDL sub-species and their function have been reported in several disease states, including atherosclerosis [4], rheumatoid arthritis (RA), systemic lupus erythematosus (SLE) [81, 82], diabetes [83], hypertension [84] and psoriasis [85–87]. Inflammation/infection triggers an APR that causes a reduction in HDL quantity and alterations in both its lipid and protein composition. Van Lenten and colleagues [88] first reported that HDL loses its ability to inhibit LDL oxidation during the APR, demonstrating that inflammation affects the structure and function of HDL.

Lipidome alterations: The phospholipid content of HDL is altered during the APR [89]. This may be due to an increase in the activity of secretory phospholipase 2 ($sPLA_2$) [90, 91]. Acute phase HDL also contains lower amounts of PE and PI along with several species of LPC with different levels of saturation. An important feature of acute phase HDL is that it contains oxidized phospholipids generated by the actions of transition metal ions, free radicals and hypochlorous acid (HOCl) [92, 93]. Formation of acute phase HDL in patients with coronary heart disease is also associated with a reduction in SM content [94]. An increase in triglycerides with a decrease in cholesteryl esters is also commonly observed in acute phase HDL [89].

Proteome alterations: Several changes in HDL-associated proteins arise in response to inflammation (**Table 2**). While a reduction in apo A-I represents perhaps the most prominent change in HDL composition, data suggest that the lipoprotein content of SAA may increase up to 1000-fold [85]. Endotoxin and inflammatory cytokines (TNF-α, IL-1β and IL-6) decrease the expression of apoA-I which leads to a decrease in circulating HDL concentration [95, 96]. In addition, an increase in the synthesis of SAA results in the displacement of apoA-I from acute phase HDL [85]. Inflammation further decreases HDL levels by inducing the upregulation of $sPLA_2$ which degrades phospholipid components of the lipoprotein particle [89]. Loss of LCAT activity [97, 98] reduces the cholesterol carrying capacity of HDL by preventing the formation of cholesterol esters. Finally, PON1 activity is reduced by inflammation in patients with RA, SLE and psoriasis and infections and is associated with a reduction in the antioxidant capacity of HDL [99–102].

Proteins[a]		**Lipids**[b]	
Increased	**Decreased**	**Increased**	**Decreased**
Serum Amyloid A (SAA)	Apo A-I	Triglycerides	Total lipid
Apo J	Apo A-II	FC	Phospholipids
$sPLA_2$	Apo C	LPC	CE
Apo E	Apo M	FFA	SM
Ceruloplasmin	LCAT		
PAF-AH	CETP		
LBP	Transferrin		
Apo A-IV	Hepatic lipase		
Apo A-V	Paraoxanase I		

[a] Adapted from Refs. [87, 96, 97, 104].
[b] Adapted from Refs. [15, 89, 94].

Table 2. Inflammation-induced changes in HDL composition.

The presence of apoM in HDL particles is thought to contribute to atheroprotection [103]. LPS and inflammatory cytokines, however, attenuate apoM mRNA levels and protein expression in Hep3B cells [104]. A decrease in serum apoM is also observed in patients with sepsis and HIV infections [104]. Further, a reduction in apoM reduces the association of S1P with HDL resulting in degradation of anti-inflammatory function [103].

The association of other apolipoproteins with HDL may impair the function of the lipoprotein. ApoO is incorporated by HDL, LDL and VLDL particles [105]. Data suggest that apoO provides structural stability for mitochondria by stabilizing the inner mitochondrial membrane and cristae [105]. Other data, however, show that overexpression of apoO degrades mitochondrial protein and increases cardiac dysfunction in hypercholesterolemic mice [106]. In cardiomyocyte cultures, upregulation of apoO was associated with an increase in ROS and apoptosis compared to control cells that were apoO-deficient [106]. ApoC is an additional, exchangeable apolipoprotein associated with HDL and apoB-containing lipoproteins. In isolated rat liver mitochondria, addition of the apoC-III isoform was shown to inhibit mitochondrial oxygen consumption and attenuate ATP formation [107]. Another study showed that enrichment of HDL with apoC-I stimulates cytochrome c release, caspase 3 cleavage and cell death in human aortic smooth muscle cells [108]. Finally, apoC-I enrichment of HDL is associated with a reduction in HDL-associated apoA-I, suggesting that loss of apoA-I and its cytoprotective effects is a component of apoC-I-mediated cell injury [107, 108]. Clearly, additional in vitro and in vivo studies are required to define the mechanistic role of specific apolipoprotein species in the development of inflammatory injury.

5. Functional consequences of acute phase HDL formation

Changes in HDL lipid and protein composition induced by the APR impair normal HDL function resulting in the formation of "dysfunctional" HDL.

Loss of cholesterol efflux ability: Since cholesterol efflux involves the participation of apoA-I, phospholipids, LCAT and CETP, several aspects of dysfunctional HDL inhibit normal reverse cholesterol transport. The reduction in apoA-I and increase in HDL-associated SAA impair cholesterol efflux capacity [109, 110]. The presence of SAA on HDL increases foam cell formation by facilitating the uptake of cholesterol esters by macrophages. At the level of the hepatocyte, this acute phase HDL impairs cholesterol uptake and degradation [111]. Decreased content of LCAT, PL and CETP on HDL also contribute to a loss of efflux activity as does the oxidative modification of apoA-I [112, 113].

Impairment of antioxidative activity: An increase in TG and decrease in cholesterol ester content in dysfunctional HDL leads to a change in conformation of the HDL particle. The formation of a TG-rich HDL particle induces structural changes in apoA-I and decreases its stability [114]. Additionally, an increase in SAA and loss of PON1 result in a reduced antioxidant capacity of the HDL particle.

Attenuation of anti-inflammatory activity: Dysfunctional HDL has an impaired capacity to counteract the action of LPS and inflammatory cytokines. The ability to regulate membrane raft cholesterol content is reduced and can thus enhance TLR activation in response to proinflammatory mediators [115]. Oxidation of apoA-I also results in a loss of functionality with respect to its ability to efflux cholesterol. The protein and lipid alterations observed (reduced apoA-I, cholesterol ester, PON1 and LCAT levels and increased TG and SAA levels) are also responsible for the attenuated anti-inflammatory activity observed with dysfunctional HDL.

6. Conclusions

HDL plays an important role in regulating atherogenesis *via* its ability to mediate reverse cholesterol transport. The ability of HDL to reduce inflammatory injury and oxidant stress has also been shown to reduce CVD risk. As discussed in this review, both protein and lipid components of the lipoprotein particle play critical roles in attenuating inflammation.

Identification of these cytoprotective HDL components has been facilitated by recent proteomic analyses. Under pathological conditions, HDL levels may be reduced and the lipoprotein may undergo biochemical and structural modification resulting in the formation of dysfunctional HDL with pro-inflammatory properties. It has been suggested that the anti-inflammatory status of HDL may be of greater predictive value for CVD risk than HDL levels per se [116, 117]. Therapeutic approaches that increase the functional properties of HDL may thus be superior to simply raising circulating HDL. Unfortunately, specific and reliable biomarkers for anti-inflammatory HDL have not been identified. Under ex vivo conditions, the quality of HDL can be assessed by studying lipoprotein effects of processes such as monocyte chemotaxis and endothelial inflammation. These assays, however, are cumbersome and time-consuming. Despite these drawbacks, there is significant interest in developing new pharmacotherapies that positively impact circulating lipoproteins. Randomized clinical trials have assessed effects of several classes of drugs on plasma cholesterol levels in patients at risk. Niacin and statins significantly lower LDL and were shown to induce modest increases in HDL [8]. Residual risk, however, may be present in patients with persistently low HDL despite a reduction in LDL. CETP inhibitors have been shown to increase HDL levels in animal models and in human subjects with low HDL [118, 119]. The ILLUMINATE trial tested effects of the CETP inhibitor torcetrapib on HDL and outcomes in high risk patients but was terminated early due to an increase in mortality due to off-target effects [7]. In ongoing studies, the antiatherogenic and anti-inflammatory effects of reconstituted HDL therapy as well as apolipoprotein mimetics are being evaluated. Recent exciting data also show that HDL serves as a carrier for functional miRNAs that suppress inflammation at the level of the endothelial cell [120]. miRNAs have also been identified that regulate HDL biogenesis [121]. These recent observations may lay the foundation for a new field of miRNA-based HDL therapeutics.

Author details

Roger White[1], Samantha Giordano[1] and Geeta Datta[2*]

*Address all correspondence to: gdatta@uabmc.edu

1 Department of Medicine, University of Alabama at Birmingham, USA

2 Division of Cardiovascular Disease, Department of Medicine, University of Alabama at Birmingham, Birmingham, AL, USA

References

[1] Gotto AM, Jr. Jeremiah Metzger Lecture: cholesterol, inflammation and atherosclerotic cardiovascular disease: is it all LDL? Transactions of the American Clinical and Climatological Association. 2011;122:256–89.

[2] Lewis B. Classification of lipoproteins and lipoprotein disorders. Journal of Clinical Pathology Supplement. 1973;5:26–31.

[3] Gordon DJ, Rifkind BM. High-density lipoprotein—the clinical implications of recent studies. The New England Journal of Medicine. 1989;321:1311–6.

[4] Gordon T, Castelli WP, Hjortland MC, Kannel WB, Dawber TR. High density lipoprotein as a protective factor against coronary heart disease. The Framingham Study. The American Journal of Medicine. 1977;62:707–14.

[5] Ridker PM, Danielson E, Fonseca FA, Genest J, Gotto AM, Jr., Kastelein JJ, et al. Rosuvastatin to prevent vascular events in men and women with elevated C-reactive protein. The New England Journal of Medicine. 2008;359:2195–207.

[6] Yusuf S, Bosch J, Dagenais G, Zhu J, Xavier D, Liu L, et al. Cholesterol lowering in intermediate-risk persons without cardiovascular disease. The New England Journal of Medicine. 2016;374:2021–31.

[7] Barter PJ, Caulfield M, Eriksson M, Grundy SM, Kastelein JJ, Komajda M, et al. Effects of torcetrapib in patients at high risk for coronary events. The New England Journal of Medicine. 2007;357:2109–22.

[8] Investigators A-H. HDL cholesterol levels receiving intensive ststin therapy. New England Journal of Medicine. 2011;365:2255–67.

[9] Schwartz GG, Olsson AG, Abt M, Ballantyne CM, Barter PJ, Brumm J, et al. Effects of dalcetrapib in patients with a recent acute coronary syndrome. The New England Journal of Medicine. 2012;367:2089–99.

[10] Plump AS, Scott CJ, Breslow JL. Human apolipoprotein A-I gene expression increases high density lipoprotein and suppresses atherosclerosis in the apolipoprotein E-deficient mouse. Proceedings of the National Academy of Sciences of the United States of America. 1994;91:9607–11.

[11] Joy T, Hegele RA. Is raising HDL a futile strategy for atheroprotection? Nature Reviews Drug discovery. 2008;7:143–55.

[12] Asztalos BF, de la Llera-Moya M, Dallal GE, Horvath KV, Schaefer EJ, Rothblat GH. Differential effects of HDL subpopulations on cellular ABCA1- and SR-BI-mediated cholesterol efflux. Journal of Lipid Research. 2005;46:2246–53.

[13] Duong PT, Weibel GL, Lund-Katz S, Rothblat GH, Phillips MC. Characterization and properties of pre beta-HDL particles formed by ABCA1-mediated cellular lipid efflux to apoA-I. Journal of Lipid Research. 2008;49:1006–14.

[14] Zhu X, Parks JS. New roles of HDL in inflammation and hematopoiesis. Annual Review of Nutrition. 2012;32:161–82.

[15] Kontush A, Lhomme M, Chapman MJ. Unraveling the complexities of the HDL lipidome. Journal of Lipid Research. 2013;54:2950–63.

[16] Nixon GF. Sphingolipids in inflammation: pathological implications and potential therapeutic targets. British Journal of Pharmacology. 2009;158:982–93.

[17] Maceyka M, Spiegel S. Sphingolipid metabolites in inflammatory disease. Nature. 2014;510:58–67.

[18] Hla T, Dannenberg AJ. Sphingolipid signaling in metabolic disorders. Cell Metabolism. 2012;16:420–34.

[19] Shah AS, Tan L, Long JL, Davidson WS. Proteomic diversity of high density lipoproteins: our emerging understanding of its importance in lipid transport and beyond. Journal of Lipid Research. 2013;54:2575–85.

[20] Vaisar T, Pennathur S, Green PS, Gharib SA, Hoofnagle AN, Cheung MC, et al. Shotgun proteomics implicates protease inhibition and complement activation in the antiinflammatory properties of HDL. The Journal of Clinical Investigation. 2007;117:746–56.

[21] Fazio S, Linton MF. Sorting out the complexities of reverse cholesterol transport: CETP polymorphisms, HDL, and coronary disease. The Journal of Clinical Endocrinology and Metabolism. 2006;91:3273–5.

[22] Du XM, Kim MJ, Hou L, Le Goff W, Chapman MJ, Van Eck M, et al. HDL particle size is a critical determinant of ABCA1-mediated macrophage cellular cholesterol export. Circulation Research. 2015;116:1133–42.

[23] Rosenson RS, Brewer HB, Jr., Davidson WS, Fayad ZA, Fuster V, Goldstein J, et al. Cholesterol efflux and atheroprotection: advancing the concept of reverse cholesterol transport. Circulation. 2012;125:1905–19.

[24] Williams KJ, Tabas I. The response-to-retention hypothesis of early atherogenesis. Arteriosclerosis, Thrombosis, and Vascular Biology. 1995;15:551–61.

[25] Gaut JP, Heinecke JW. Mechanisms for oxidizing low-density lipoprotein. Insights from patterns of oxidation products in the artery wall and from mouse models of atherosclerosis. Trends in Cardiovascular Medicine. 2001;11:103–12.

[26] Kontush A, Chapman MJ. Antiatherogenic function of HDL particle subpopulations: focus on antioxidative activities. Current Opinion in Lipidology. 2010;21:312–8.

[27] Zerrad-Saadi A, Therond P, Chantepie S, Couturier M, Rye KA, Chapman MJ, et al. HDL3-mediated inactivation of LDL-associated phospholipid hydroperoxides is determined by the redox status of apolipoprotein A-I and HDL particle surface lipid rigidity: relevance to inflammation and atherogenesis. Arteriosclerosis, Thrombosis, and Vascular Biology. 2009;29:2169–75.

[28] Garner B, Waldeck AR, Witting PK, Rye KA, Stocker R. Oxidation of high density lipoproteins. II. Evidence for direct reduction of lipid hydroperoxides by methionine residues of apolipoproteins AI and AII. The Journal of Biological Chemistry. 1998;273:6088–95.

[29] Navab M, Hama SY, Cooke CJ, Anantharamaiah GM, Chaddha M, Jin L, et al. Normal high density lipoprotein inhibits three steps in the formation of mildly oxidized low density lipoprotein: step 1. Journal of Lipid Research. 2000;41:1481–94.

[30] Davidson WS, Silva RA, Chantepie S, Lagor WR, Chapman MJ, Kontush A. Proteomic analysis of defined HDL subpopulations reveals particle-specific protein clusters: relevance to antioxidative function. Arteriosclerosis, Thrombosis, and Vascular Biology. 2009;29:870–6.

[31] Kontush A, Chantepie S, Chapman MJ. Small, dense HDL particles exert potent protection of atherogenic LDL against oxidative stress. Arteriosclerosis, Thrombosis, and Vascular Biology. 2003;23:1881–8.

[32] Kontush A, de Faria EC, Chantepie S, Chapman MJ. Antioxidative activity of HDL particle subspecies is impaired in hyperalphalipoproteinemia: relevance of enzymatic and physicochemical properties. Arteriosclerosis, Thrombosis, and Vascular Biology. 2004;24:526–33.

[33] Frostegard J. Immunity, atherosclerosis and cardiovascular disease. BMC Medicine. 2013;11:117.

[34] Tabas I, Garcia-Cardena G, Owens GK. Recent insights into the cellular biology of atherosclerosis. The Journal of Cell Biology. 2015;209:13–22.

[35] Manduteanu I, Simionescu M. Inflammation in atherosclerosis: a cause or a result of vascular disorders? Journal of Cellular and Molecular Medicine. 2012;16:1978–90.

[36] Catapano AL, Pirillo A, Bonacina F, Norata GD. HDL in innate and adaptive immunity. Cardiovascular Research. 2014;103:372–83.

[37] Guo L, Ai J, Zheng Z, Howatt DA, Daugherty A, Huang B, et al. High density lipoprotein protects against polymicrobe-induced sepsis in mice. The Journal of Biological Chemistry. 2013;288:17947–53.

[38] Levine DM, Parker TS, Donnelly TM, Walsh A, Rubin AL. In vivo protection against endotoxin by plasma high density lipoprotein. Proceedings of the National Academy of Sciences of the United States of America. 1993;90:12040–4.

[39] Morin EE, Guo L, Schwendeman A, Li XA. HDL in sepsis—risk factor and therapeutic approach. Frontiers in Pharmacology. 2015;6:244.

[40] White CR, Smythies LE, Crossman DK, Palgunachari MN, Anantharamaiah GM, Datta G. Regulation of pattern recognition receptors by the apolipoprotein A-I mimetic peptide 4F. Arteriosclerosis, Thrombosis, and Vascular Biology. 2012;32:2631–9.

[41] Triantafilou M, Manukyan M, Mackie A, Morath S, Hartung T, Heine H, et al. Lipoteichoic acid and toll-like receptor 2 internalization and targeting to the Golgi are lipid raft-dependent. The Journal of Biological Chemistry. 2004;279:40882–9.

[42] Labonte AC, Tosello-Trampont AC, Hahn YS. The role of macrophage polarization in infectious and inflammatory diseases. Molecules and Cells. 2014;37:275–85.

[43] Lawrence T, Natoli G. Transcriptional regulation of macrophage polarization: enabling diversity with identity. Nature Reviews Immunology. 2011;11:750–61.

[44] Martinez FO, Sica A, Mantovani A, Locati M. Macrophage activation and polarization. Frontiers in Bioscience: a Journal and Virtual Library. 2008;13:453–61.

[45] Mosser DM, Edwards JP. Exploring the full spectrum of macrophage activation. Nature Reviews Immunology. 2008;8:958–69.

[46] Mosser DM. The many faces of macrophage activation. Journal of Leukocyte Biology. 2003;73:209–12.

[47] Porta C, Rimoldi M, Raes G, Brys L, Ghezzi P, Di Liberto D, et al. Tolerance and M2 (alternative) macrophage polarization are related processes orchestrated by p50 nuclear factor kappaB. Proceedings of the National Academy of Sciences of the United States of America. 2009;106:14978–83.

[48] Smythies LE, White CR, Maheshwari A, Palgunachari MN, Anantharamaiah GM, Chaddha M, et al. Apolipoprotein A-I mimetic 4F alters the function of human monocyte-derived macrophages. American Journal of Physiology Cell Physiology. 2010;298:C1538–48.

[49] Feig JE, Rong JX, Shamir R, Sanson M, Vengrenyuk Y, Liu J, et al. HDL promotes rapid atherosclerosis regression in mice and alters inflammatory properties of plaque monocyte-derived cells. Proceedings of the National Academy of Sciences of the United States of America. 2011;108:7166–71.

[50] Argraves KM, Gazzolo PJ, Groh EM, Wilkerson BA, Matsuura BS, Twal WO, et al. High density lipoprotein-associated sphingosine 1-phosphate promotes endothelial barrier function. The Journal of Biological Chemistry. 2008;283:25074–81.

[51] Keul P, Sattler K, Levkau B. HDL and its sphingosine-1-phosphate content in cardioprotection. Heart Failure Reviews. 2007;12:301–6.

[52] Theilmeier G, Schmidt C, Herrmann J, Keul P, Schafers M, Herrgott I, et al. High-density lipoproteins and their constituent, sphingosine-1-phosphate, directly protect the heart against ischemia/reperfusion injury in vivo via the S1P3 lysophospholipid receptor. Circulation. 2006;114:1403–9.

[53] Hughes JE, Srinivasan S, Lynch KR, Proia RL, Ferdek P, Hedrick CC. Sphingosine-1-phosphate induces an antiinflammatory phenotype in macrophages. Circulation Research. 2008;102:950–8.

[54] Amo T, Yadava N, Oh R, Nicholls DG, Brand MD. Experimental assessment of bioenergetic differences caused by the common European mitochondrial DNA haplogroups H and T. Gene. 2008;411:69–76.

[55] Fetterman JL, Zelickson BR, Johnson LW, Moellering DR, Westbrook DG, Pompilius M, et al. Mitochondrial genetic background modulates bioenergetics and susceptibility to acute cardiac volume overload. The Biochemical Journal. 2013;455:157–67.

[56] Lee J, Giordano S, Zhang J. Autophagy, mitochondria and oxidative stress: cross-talk and redox signalling. The Biochemical Journal. 2012;441:523–40.

[57] Garcia-Heredia A, Marsillach J, Rull A, Triguero I, Fort I, Mackness B, et al. Paraoxonase-1 inhibits oxidized low-density lipoprotein-induced metabolic alterations and apoptosis in endothelial cells: a nondirected metabolomic study. Mediators of Inflammation. 2013;2013:156053.

[58] Sangle GV, Chowdhury SK, Xie X, Stelmack GL, Halayko AJ, Shen GX. Impairment of mitochondrial respiratory chain activity in aortic endothelial cells induced by glycated low-density lipoprotein. Free Radical Biology & Medicine. 2010;48:781–90.

[59] Dadabayev AR, Yin G, Latchoumycandane C, McIntyre TM, Lesnefsky EJ, Penn MS. Apolipoprotein A1 regulates coenzyme Q10 absorption, mitochondrial function, and infarct size in a mouse model of myocardial infarction. The Journal of Nutrition. 2014;144:1030–6.

[60] Kalakech H, Hibert P, Prunier-Mirebeau D, Tamareille S, Letournel F, Macchi L, et al. RISK and SAFE signaling pathway involvement in apolipoprotein A-I-induced cardioprotection. PLoS One. 2014;9:e107950.

[61] Lecour S. Multiple protective pathways against reperfusion injury: a SAFE path without Aktion? Journal of Molecular and Cellular Cardiology. 2009;46:607–9.

[62] Szczepanek K, Chen Q, Derecka M, Salloum FN, Zhang Q, Szelag M, et al. Mitochondrial-targeted signal transducer and activator of transcription 3 (STAT3) protects against ischemia-induced changes in the electron transport chain and the generation of reactive oxygen species. The Journal of Biological Chemistry. 2011;286:29610–20.

[63] Tammineni P, Anugula C, Mohammed F, Anjaneyulu M, Larner AC, Sepuri NB. The import of the transcription factor STAT3 into mitochondria depends on GRIM-19, a component of the electron transport chain. The Journal of Biological Chemistry. 2013;288:4723–32.

[64] Wegrzyn J, Potla R, Chwae YJ, Sepuri NB, Zhang Q, Koeck T, et al. Function of mitochondrial Stat3 in cellular respiration. Science. 2009;323:793–7.

[65] Frias MA, Lecour S, James RW, Pedretti S. High density lipoprotein/sphingosine-1-phosphate-induced cardioprotection: role of STAT3 as part of the SAFE pathway. Jak-Stat. 2012;1:92–100.

[66] Somers SJ, Frias M, Lacerda L, Opie LH, Lecour S. Interplay between SAFE and RISK pathways in sphingosine-1-phosphate-induced cardioprotection. Cardiovascular Drugs and Therapy/sponsored by the International Society of Cardiovascular Pharmacotherapy. 2012;26:227–37.

[67] Zhang J, Honbo N, Goetzl EJ, Chatterjee K, Karliner JS, Gray MO. Signals from type 1 sphingosine 1-phosphate receptors enhance adult mouse cardiac myocyte survival during hypoxia. American Journal of Physiology Heart and Circulatory Physiology. 2007;293:H3150–8.

[68] Gomez L, Paillard M, Price M, Chen Q, Teixeira G, Spiegel S, et al. A novel role for mitochondrial sphingosine-1-phosphate produced by sphingosine kinase-2 in PTP-mediated cell survival during cardioprotection. Basic Research in Cardiology. 2011;106:1341–53.

[69] Merkwirth C, Langer T. Prohibitin function within mitochondria: essential roles for cell proliferation and cristae morphogenesis. Biochimica et biophysica acta. 2009; 1793:27–32.

[70] Strub GM, Paillard M, Liang J, Gomez L, Allegood JC, Hait NC, et al. Sphingosine-1-phosphate produced by sphingosine kinase 2 in mitochondria interacts with prohibitin 2 to regulate complex IV assembly and respiration. FASEB Journal. 2011;25:600–12.

[71] de Silva HV, Stuart WD, Duvic CR, Wetterau JR, Ray MJ, Ferguson DG, et al. A 70-kDa apolipoprotein designated ApoJ is a marker for subclasses of human plasma high density lipoproteins. The Journal of Biological Chemistry. 1990;265:13240–7.

[72] Park S, Mathis KW, Lee IK. The physiological roles of apolipoprotein J/clusterin in metabolic and cardiovascular diseases. Reviews in Endocrine & Metabolic Disorders. 2014;15:45–53.

[73] Trougakos IP. The molecular chaperone apolipoprotein J/clusterin as a sensor of oxidative stress: implications in therapeutic approaches—a mini-review. Gerontology. 2013;59:514–23.

[74] Jun HO, Kim DH, Lee SW, Lee HS, Seo JH, Kim JH, et al. Clusterin protects H9c2 cardiomyocytes from oxidative stress-induced apoptosis via Akt/GSK-3beta signaling pathway. Experimental & Molecular Medicine. 2011;43:53–61.

[75] Blaho VA, Hla T. An update on the biology of sphingosine 1-phosphate receptors. Journal of Lipid Research. 2014;55:1596–608.

[76] Christoffersen C, Ahnstrom J, Axler O, Christensen EI, Dahlback B, Nielsen LB. The signal peptide anchors apolipoprotein M in plasma lipoproteins and prevents rapid clearance of apolipoprotein M from plasma. The Journal of Biological Chemistry. 2008;283:18765–72.

[77] Dahlback B, Nielsen LB. Apolipoprotein M affecting lipid metabolism or just catching a ride with lipoproteins in the circulation? Cellular and Molecular Life Sciences: CMLS. 2009;66:559–64.

[78] Elsoe S, Christoffersen C, Luchoomun J, Turner S, Nielsen LB. Apolipoprotein M promotes mobilization of cellular cholesterol in vivo. Biochimica et Biophysica Acta. 2013;1831:1287–92.

[79] Christoffersen C, Obinata H, Kumaraswamy SB, Galvani S, Ahnstrom J, Sevvana M, et al. Endothelium-protective sphingosine-1-phosphate provided by HDL-associated apolipoprotein M. Proceedings of the National Academy of Sciences of the United States of America. 2011;108:9613–8.

[80] Duan J, Dahlback B, Villoutreix BO. Proposed lipocalin fold for apolipoprotein M based on bioinformatics and site-directed mutagenesis. FEBS Letters. 2001;499:127–32.

[81] Ormseth MJ, Stein CM. High-density lipoprotein function in rheumatoid arthritis. Current Opinion in Lipidology. 2016;27:67–75.

[82] Hahn BH, Grossman J, Ansell BJ, Skaggs BJ, McMahon M. Altered lipoprotein metabolism in chronic inflammatory states: proinflammatory high-density lipoprotein and accelerated atherosclerosis in systemic lupus erythematosus and rheumatoid arthritis. Arthritis Research & Therapy. 2008;10:213.

[83] Van Linthout S, Spillmann F, Schultheiss HP, Tschope C. High-density lipoprotein at the interface of type 2 diabetes mellitus and cardiovascular disorders. Current Pharmaceutical Design. 2010;16:1504–16.

[84] Odden MC, Tager IB, Gansevoort RT, Bakker SJ, Fried LF, Newman AB, et al. Hypertension and low HDL cholesterol were associated with reduced kidney function across the age spectrum: a collaborative study. Annals of Epidemiology. 2013;23:106–11.

[85] Feingold KR, Grunfeld C. Effect of inflammation on HDL structure and function. Current Opinion in Lipidology. 2016;27:521–30.

[86] Holzer M, Wolf P, Curcic S, Birner-Gruenberger R, Weger W, Inzinger M, et al. Psoriasis alters HDL composition and cholesterol efflux capacity. Journal of Lipid Research. 2012;53:1618–24.

[87] Mehta NN, Gelfand JM. High-density lipoprotein cholesterol function improves after successful treatment of psoriasis: a step forward in the right direction. The Journal of Investigative Dermatology. 2014;134:592–5.

[88] Van Lenten BJ, Hama SY, de Beer FC, Stafforini DM, McIntyre TM, Prescott SM, et al. Anti-inflammatory HDL becomes pro-inflammatory during the acute phase response. Loss of protective effect of HDL against LDL oxidation in aortic wall cell cocultures. The Journal of Clinical Investigation. 1995;96:2758–67.

[89] Khovidhunkit W, Kim MS, Memon RA, Shigenaga JK, Moser AH, Feingold KR, et al. Effects of infection and inflammation on lipid and lipoprotein metabolism: mechanisms and consequences to the host. Journal of Lipid Research. 2004;45:1169–96.

[90] Crowl RM, Stoller TJ, Conroy RR, Stoner CR. Induction of phospholipase A2 gene expression in human hepatoma cells by mediators of the acute phase response. The Journal of Biological Chemistry. 1991;266:2647–51.

[91] Pruzanski W, Vadas P, Browning J. Secretory non-pancreatic group II phospholipase A2: role in physiologic and inflammatory processes. Journal of Lipid Mediators. 1993;8:161–7.

[92] Navab M, Berliner JA, Subbanagounder G, Hama S, Lusis AJ, Castellani LW, et al. HDL and the inflammatory response induced by LDL-derived oxidized phospholipids. Arteriosclerosis, Thrombosis, and Vascular Biology. 2001;21:481–8.

[93] Stocker R. Lipoprotein oxidation: mechanistic aspects, methodological approaches and clinical relevance. Current Opinion in Lipidology. 1994;5:422–33.

[94] Pruzanski W, Stefanski E, de Beer FC, de Beer MC, Ravandi A, Kuksis A. Comparative analysis of lipid composition of normal and acute-phase high density lipoproteins. Journal of Lipid Research. 2000;41:1035–47.

[95] Ettinger WH, Varma VK, Sorci-Thomas M, Parks JS, Sigmon RC, Smith TK, et al. Cytokines decrease apolipoprotein accumulation in medium from Hep G2 cells. Arteriosclerosis and Thrombosis: A Journal of Vascular Biology/American Heart Association. 1994;14:8–13.

[96] Haas MJ, Horani M, Mreyoud A, Plummer B, Wong NC, Mooradian AD. Suppression of apolipoprotein AI gene expression in HepG2 cells by TNF alpha and IL-1beta. Biochimica et Biophysica Acta. 2003;1623:120–8.

[97] Navab M, Hama-Levy S, Van Lenten BJ, Fonarow GC, Cardinez CJ, Castellani LW, et al. Mildly oxidized LDL induces an increased apolipoprotein J/paraoxonase ratio. The Journal of Clinical Investigation. 1997;99:2005–19.

[98] Van Lenten BJ, Wagner AC, Nayak DP, Hama S, Navab M, Fogelman AM. High-density lipoprotein loses its anti-inflammatory properties during acute influenza a infection. Circulation. 2001;103:2283–8.

[99] Draganov D, Teiber J, Watson C, Bisgaier C, Nemzek J, Remick D, et al. PON1 and oxidative stress in human sepsis and an animal model of sepsis. Advances in Experimental Medicine and Biology. 2010;660:89–97.

[100] He L, Qin S, Dang L, Song G, Yao S, Yang N, et al. Psoriasis decreases the anti-oxidation and anti-inflammation properties of high-density lipoprotein. Biochimica et Biophysica Acta. 2014;1841:1709–15.

[101] Isik A, Koca SS, Ustundag B, Celik H, Yildirim A. Paraoxonase and arylesterase levels in rheumatoid arthritis. Clinical Rheumatology. 2007;26:342–8.

[102] Novak F, Vavrova L, Kodydkova J, Novak F, Sr., Hynkova M, Zak A, et al. Decreased paraoxonase activity in critically ill patients with sepsis. Clinical and Experimental Medicine. 2010;10:21–5.

[103] Ren K, Tang ZL, Jiang Y, Tan YM, Yi GH. Apolipoprotein M. Clinica Chimica Acta; International Journal of Clinical Chemistry. 2015;446:21–9.

[104] Feingold KR, Shigenaga JK, Chui LG, Moser A, Khovidhunkit W, Grunfeld C. Infection and inflammation decrease apolipoprotein M expression. Atherosclerosis. 2008;199:19–26.

[105] Lamant M, Smih F, Harmancey R, Philip-Couderc P, Pathak A, Roncalli J, et al. ApoO, a novel apolipoprotein, is an original glycoprotein up-regulated by diabetes in human heart. The Journal of Biological Chemistry. 2006;281:36289–302.

[106] Turkieh A, Caubere C, Barutaut M, Desmoulin F, Harmancey R, Galinier M, et al. Apolipoprotein O is mitochondrial and promotes lipotoxicity in heart. The Journal of Clinical Investigation. 2014;124:2277–86.

[107] Panin LE, Shalbueva NI, Polyakov LM. Effects of apolipoproteins C on oxidative phosphorylation in rat liver mitochondria. Bulletin of Experimental Biology and Medicine. 2000;130:769–71.

[108] Kolmakova A, Kwiterovich P, Virgil D, Alaupovic P, Knight-Gibson C, Martin SF, et al. Apolipoprotein C-I induces apoptosis in human aortic smooth muscle cells via recruiting neutral sphingomyelinase. Arteriosclerosis, Thrombosis, and Vascular Biology. 2004;24:264–9.

[109] Banka CL, Yuan T, de Beer MC, Kindy M, Curtiss LK, de Beer FC. Serum amyloid A (SAA): influence on HDL-mediated cellular cholesterol efflux. Journal of Lipid Research. 1995;36:1058–65.

[110] Kontush A, Chapman MJ. Functionally defective high-density lipoprotein: a new therapeutic target at the crossroads of dyslipidemia, inflammation, and atherosclerosis. Pharmacological Reviews. 2006;58:342–74.

[111] Cai L, de Beer MC, de Beer FC, van der Westhuyzen DR. Serum amyloid A is a ligand for scavenger receptor class B type I and inhibits high density lipoprotein binding and selective lipid uptake. The Journal of Biological Chemistry. 2005;280:2954–61.

[112] Shao B, Bergt C, Fu X, Green P, Voss JC, Oda MN, et al. Tyrosine 192 in apolipoprotein A-I is the major site of nitration and chlorination by myeloperoxidase, but only chlorination markedly impairs ABCA1-dependent cholesterol transport. The Journal of Biological Chemistry. 2005;280:5983–93.

[113] Zheng L, Settle M, Brubaker G, Schmitt D, Hazen SL, Smith JD, et al. Localization of nitration and chlorination sites on apolipoprotein A-I catalyzed by myeloperoxidase in human atheroma and associated oxidative impairment in ABCA1-dependent cholesterol efflux from macrophages. The Journal of Biological Chemistry. 2005;280:38–47.

[114] Curtiss LK, Bonnet DJ, Rye KA. The conformation of apolipoprotein A-I in high-density lipoproteins is influenced by core lipid composition and particle size: a surface plasmon resonance study. Biochemistry. 2000;39:5712–21.

[115] Azzam KM, Fessler MB. Crosstalk between reverse cholesterol transport and innate immunity. Trends in Endocrinology and Metabolism: TEM. 2012;23:169–78.

[116] deGoma EM, deGoma RL, Rader DJ. Beyond high-density lipoprotein cholesterol levels evaluating high-density lipoprotein function as influenced by novel therapeutic approaches. Journal of the American College of Cardiology. 2008;51:2199–211.

[117] Luscher TF, Landmesser U, von Eckardstein A, Fogelman AM. High-density lipoprotein: vascular protective effects, dysfunction, and potential as therapeutic target. Circulation Research. 2014;114:171–82.

[118] Brousseau ME, Schaefer EJ, Wolfe ML, Bloedon LT, Digenio AG, Clark RW, et al. Effects of an inhibitor of cholesteryl ester transfer protein on HDL cholesterol. The New England Journal of Medicine. 2004;350:1505–15.

[119] Morehouse LA, Sugarman ED, Bourassa PA, Sand TM, Zimetti F, Gao F, et al. Inhibition of CETP activity by torcetrapib reduces susceptibility to diet-induced atherosclerosis in New Zealand White rabbits. Journal of Lipid Research. 2007;48:1263–72.

[120] Tabet F, Vickers KC, Cuesta Torres LF, Wiese CB, Shoucri BM, Lambert G, et al. HDL-transferred microRNA-223 regulates ICAM-1 expression in endothelial cells. Nature Communications. 2014;5:3292.

[121] Rayner KJ, Moore KJ. MicroRNA control of high-density lipoprotein metabolism and function. Circulation Research. 2014;114:183–92.

Chapter 5

Plasma Lipoproteins as Crucial Components of Host Defence Against Infections

Kaustubh Bora and Probodh Borah

Additional information is available at the end of the chapter

http://dx.doi.org/10.5772/67601

Abstract

Interactions between lipoproteins and infectious microorganisms are diverse and often multifaceted. There is a growing body of evidence which suggests that circulating plasma lipoproteins play an important role in warding off various infections. They are increasingly recognized as vital components of the host immune system. The purpose of this chapter is to provide the reader with an overview of this emerging domain. We review the anti-infective role of different lipoprotein particles and their components and further highlight the known molecular mechanisms involved therein. Instances where lipoproteins facilitate infections instead of protecting against them are also summarized. Finally, broad implications for the future in this active line of research are discussed.

Keywords: lipoproteins, apolipoproteins, lipids, infection, immune system

1. Introduction

Circulating lipoproteins are macromolecular complexes of lipids and specific proteins (known as apolipoproteins). They facilitate the transport and distribution of various lipids (such as cholesterol, cholesteryl esters, triglycerides, and phospholipids) via blood throughout the body. Owing to their hydrophobicity, they are otherwise sparingly soluble in the predominantly aqueous plasma [1]. Scientific work on plasma lipoproteins has historically focused on their role in atherosclerotic changes and cardiovascular health. Much of the impetus in this line of enquiry was provided by the Framingham Heart Study (FHS) that was started in 1948 by the National Heart Institute (NHI). The FHS and a number of large clinico-epidemiological studies thereafter have been instrumental in advancing our knowledge about the link between circulating lipoproteins and cardiovascular health [1–5]. There is a growing body of evidence suggesting that plasma lipoproteins are crucial players in a host of other conditions as well, viz. neurodegeneration [6], psychiatric ailments [7], and various cancers [8, 9], to name a few.

Although the earliest reports about the relationship of lipids and lipoproteins with various infections date back to 1940s and 1950s [10–12], yet the interest on lipoproteins for a long time was mostly revolved around noncommunicable disorders. However, in a marked departure from this conventional outlook, the importance of circulating lipoproteins in relation to infectious diseases is now widely recognized. Perhaps the best example in this regard is the study of the role of high-density lipoproteins (HDL) particles in conferring immunity against *Trypanosoma brucei brucei* in humans [13]. This change in the outlook is probably due to the fact that derangements (quantitative as well as qualitative) in plasma lipoproteins that were earlier documented in a variety of infections, viz. bacterial, viral, and parasitic [14–16] have now been corroborated by experimental evidence as well [17–19], such that there is an improved understanding of the underlying mechanisms at a molecular level.

Lipoproteins represent structurally and functionally a very diverse species of complex particles with dynamic interactions that travel throughout the body through circulation. Thus, they are increasingly appreciated as components of the innate immune system [15, 17, 20]. Recent evidence suggests that lipoproteins are also involved in adaptive immune responses [21]. On the basis of difference in hydrated densities (as determined by their rate in sedimentation on ultracentrifugation in salt solutions), human plasma proteins have been traditionally divided into four major groups—chylomicrons, very low-density lipoproteins (VLDL), low-density lipoproteins (LDL), and high-density lipoproteins (HDL) [1, 4, 5]. Apart from these four major groups, sometimes, other lipoprotein classes are also described, such as intermediate-density lipoproteins (IDL)—produced by catabolism of VLDL, and lipoprotein(a) [Lp(a)]—which is structurally similar to LDL and has a density range that overlaps that of HDL [1, 4, 5]. Many of these lipoprotein particles or their components (e.g., lipids or apolipoproteins) have been found to exert anti-infective role.

This chapter aims to review this emerging domain where plasma lipoproteins are now widely recognized as important players of the host immune system. We have summarized the different types of lipoprotein derangements during various infections, the anti-infective role of lipoproteins in conferring protection against pathogens, and the known molecular mechanisms involved therein.

2. Lipoproteins in relation to various infections

Lipoprotein derangements and infections appear to have a bidirectional relationship. This means that alterations in circulatory lipoproteins can modulate or predispose to infections, and conversely, alterations in circulating lipoproteins can be an outcome of the infections themselves. In other words, lipoprotein derangements can both be a contributory cause and a resultant effect of infections. The focus of this chapter is predominantly on the former relationship which underscores the role of lipoproteins in modulating susceptibility to infections.

http://dx.doi.org/10.5772/67601

The latter relationship is linked to the active phase of the infection and has been recorded in relation to several kinds of infectious agents [10–12, 22–24]. Such lipoprotein derangements are a part of the acute-phase responses (APR) mounted by the body and are beyond the scope of this chapter. Similarly, lipoprotein derangements can occur as a result of drug therapy against infections (e.g., dyslipidemia in HIV-AIDS patients due to anti-retroviral therapy) [25–28] and are outside the purview of this chapter too.

Generally speaking, most of the experimental and clinical evidences suggest that high levels of lipoproteins and lipids are protective against respiratory and gastrointestinal infections [29–31]. Case studies in homogenous populations residing in high-infection environments affirm this view. For instance, the Tsimane people of Bolivian Amazon have very high burden of infection, and this is often attributed to the low levels of lipids and lipoproteins [32]. Likewise, in the Shipibo people, another indigenous group in the Amazon, the density of parasitic infection correlates inversely to the HDL levels [33]. Further, reduced levels of apolipoproteins in hospital-based studies have been reported to be associated with increased susceptibility to nosocomial infections following severe trauma [34]. However, generalizations are difficult and exceptions to these trends have also been reported [28], and the mechanism involved is not clear.

In the account that follows, we give an overview of different infections where lipoproteins provide protection.

2.1. Viral infections

Lipoproteins, particularly HDL particles, have been found to account for part of the broad nonspecific antiviral activity of human serum [35, 36]. Such antiviral activity of lipoproteins has been detected across a wide spectrum of enveloped as well as nonenveloped DNA and RNA viruses. Examples include Rabies virus, Rubella virus, Japanese encephalitis virus (JEV), Poliovirus, Epstein-Barr virus (EBV), Herpes simplex virus (HSV), Vaccinia virus, New Castle disease virus, and Vesicular stomatitis virus (VSV), to name a few [35, 37–41]. This is in tune with the protection conferred by other components of the innate immune system, which are often nonspecific and broad-based. However, some lipoproteins (e.g., LDL and VLDL) have been found to be particularly active against certain viruses (e.g., JEV, Rubella, Rabies, and VSV) [35, 38, 40].

2.2. Bacterial infections

Lipoproteins are protective against several toxins produced by pathogenic bacteria. Lipoproteins can neutralize lipopolysaccharides (LPS) from Gram-negative bacteria [30, 36, 42]. LPS are implicated in complications of Gram-negative bacteraemia such as endotoxic shock and disseminated intravascular coagulation. Several classes of lipoproteins, such as LDL, VLDL, HDL, Lp(a), and chylomicrons, can potentially help in neutralizing LPS [30, 36, 43–45]. In fact, infusion of reconstituted HDL particles (rHDL) has been shown to protect against Gram-negative bacteraemia and endotoxic shock and to further blunt the LPS-induced unregulated activation of the coagulation cascade [46–48].

Lipoproteins are protective against Gram-positive organisms too. Lipoproteins have been shown to inactivate lipotechoic acid (LTA) and alpha-toxin from Gram-positive bacteria such as *Staphylococcus aureus* [30, 36, 49].

In addition to these toxin-neutralizing effects, lipoproteins can directly interact with cell surface virulence factors in bacteria and help in limiting their pathogenicity. Such interactions have been noted in infections by *Yersinia pestis* [50] and many Group A *Streptococcus* (GAS) strains [30, 51]. Besides, experiments in knockout animals have revealed that apoE-/- mice are susceptible to infection by *Listeria monocytogenes* and *Mycobacterium tuberculosis* [52, 53].

2.3. Parasitic infections

Humans are immune to infection by the parasite *Trypanosoma brucei brucei*. This protection is attributed to a subset of HDL particles called trypanosome lytic factors (TLFs), present in human serum [13]. TLFs have also been shown to ameliorate infection by *Leishmania* sp. [15]. However, this protection does not extend against other trypanosomes such as *Trypanosoma cruzi, Trypanosoma brucei rhodesiense,* and *Trypanosoma brucei gambiense* [13]. Lipoproteins are suggested to modulate the infectivity of malaria parasite and *Schistosoma* as well [36, 54–56].

3. Anti-infective mechanisms of lipoproteins

The biological mechanisms for the anti-infective role of lipoproteins are diverse. While, in some instances, the complete ensemble of a lipoprotein particle has been found to contribute to the immunological defenses, in some other occasions the individual constituents (such as apoproteins or lipid moieties such as phospholipids and cholesterol) are credited to be involved [36, 41, 57]. A broad scheme of the anti-infective mechanisms with respect to circulating lipoproteins is provided below. Experimental evidences from in vivo and in vitro studies suggest that these schemes are actually recurring themes. These strategies are common to a variety of antimicrobial defenses mobilized by circulating lipoproteins against a plethora of infectious agents.

3.1. Inhibiting the entry of intracellular pathogens into host cells

Lipoproteins can inhibit the attachment and subsequent entry of pathogens into their target cells. This defensive mechanism of lipoproteins has been particularly well described in relation to viruses. The presence of HDL is capable of retaining viruses on the cell surface, lending credibility to this idea. Apoproteins (such as apoA-I) in host circulatory lipoproteins contain stretches of amphipathic residues that have been proposed to interact with amphipathic counterparts in alpha-helices of viral envelope glycoproteins. These interactions interfere with membrane fusion and entry of viruses into host cells. Synthetic amphipathic peptide analogues of apoA-I can also exert similar effects [36, 58, 59]. In fact, such analogues have been found to inhibit HIV-induced syncytium formation [60]. Inhibition of viral penetration inside host cells is also supported by VLDL. Recent in vivo studies have revealed that VLDL in serum effectively blocks hepatitis C virus (HCV) cell attachment, thereby acting as a restriction factor against HCV infection [18]. Clinical studies have earlier revealed that serum level of apoC-III (an integral apolipoprotein in VLDL) was a significant predictor of chronic hepatitis C infection and associated hepatic fibrosis [16].

Alternatively, viral infection stimulates production of interferons, which in turn induce secretion of some soluble forms of lipoprotein receptors. These soluble receptors can modulate viral pathogenesis. For instance, a soluble LDL receptor shed during hepatitis and rhinovirus infections is used by the viruses for gaining entry into their respective host target cells [36]. However, endogenous LDL competes with these viruses or such similar viruses for cellular uptake, protecting the host against infection. Such receptors are also implicated in virus assembly and budding [61, 62]. Likewise, a VLDL receptor fragment that binds rhinoviruses has also been described in cell culture studies [63].

In addition to viruses, circulating lipoproteins have been found to prevent the entry of non-viral intracellular pathogens. For example, lipoproteins can interfere with the adhesion of *Salmonella typhimurium* to host cells and subsequent organ invasion [64].

3.2. Inactivating the effect of microbial toxins

Lipoproteins effectively neutralize bacterial toxins such as LPS (from Gram-negative bacteria) and LTA (from Gram-positive bacteria) and enhance their clearance. The mechanisms involved in inactivating LPS are particularly well established [30, 36, 42–45, 65, 66]. The lipid components of lipoproteins are vital in this regard. Ultrastructural studies have shown that LPS binding with LDL causes fatty acyl chain of crucial lipid moieties in LPS to be incorporated into the phospholipid surface of lipoproteins. This masks the active sites of LPS and attenuates their toxic action [36, 67].

Binding with lipoproteins also enhances the clearance of LPS. During Gram-negative bacteremia, LPS released in the circulation is primarily taken up by macrophages in liver (Kupffer cells). The macrophages thus activated cause a splurge of pro-inflammatory cytokines, which are responsible for the LPS-induced septic shock. However, binding of LPS with lipoproteins prevents this and causes two-pronged benefits. Firstly, on binding with lipoproteins, the uptake of LPS by hepatic macrophages decreases, which prevents their activation and cytokine release [36, 68–70]. Lipoproteins can prevent the LPS-mediated activation and release of cytokines from peripheral monocytes/macrophages too. Lipoproteins have been found to promote the release of LPS from the cell surface of monocytes to which they were bound, further dampening the cellular response [36, 71]. Secondly, the lipoprotein bound LPS are instead taken up by hepatocytes that lead to their rapid secretion into bile [36, 68–70]. Triglyceride-rich lipoproteins such as chylomicrons and VLDL are especially active in accelerating the clearance of LPS in this fashion [36, 68].

In a somewhat analogous manner, lipoproteins are believed to neutralize the toxic effects of LTA [72]. Further, potent peptide toxins such as phenol-soluble modulins (PSM) secreted by bacteria such as *Staphylococcus aureus* can also be inactivated by lipoproteins such as HDL, LDL, and VLDL. Highest binding and neutralizing potentials of *Staphylococcal* PSMs are displayed by HDL [17].

3.3. Lysis of pathogens

Certain pathogens are directly lysed by plasma lipoproteins or their components. A good example of this is the lysis of the parasite *Trypanosoma brucei brucei* [13, 36]. This lipoprotein-mediated lysis is attributed to two distinct trypanosome lytic factors (TLFs), namely TLF1 and TLF2. TLF1 is actually a lipid-rich subset of HDL that contains mostly apoA-I and haptoglobulin-related protein (HRP) with some amount of other proteins such as apoA-II, apoL-I, and paraoxonase. On the other hand, TLF2 is lipid-poor lipoprotein complex that contains apoA-I, HRP, and immunoglobulin M [73–75]. It is believed that apoL-I and HRP in TLFs target the parasites within the acidic parasitophorous vacuoles of macrophages and damage them directly without taking recourse to macrophage activation [15]. It is noteworthy that *Trypanosoma cruzi*, a trypanosome to which humans are susceptible, cleaves apoA-I, the chief protein constituent of HDL using cruzipain, a cysteine protease present in the cell membrane as well as internal lysosomal structure of the parasite [76]. Such targeted breakdown of vital lipoprotein constituents may aid the *Trypanosoma cruzi* parasite in evading the anti-parasitic action of TLFs.

3.4. Promoting opsonization

Experiments involving in vitro and ex vivo systems have suggested that some lipoproteins such as LDL may act as opsonins and enhance phagocytosis of several types of Group A *Streptococcus* (GAS) bacteria by monocytes. Interaction of LDL with CD36 scavenger receptor expressed in monocytes and streptococcal collagen such as protein 1 (Scl1) present on the cell surface of GAS is believed to underlie this phagocytosis promoting activity [19].

3.5. Activation of complement system

Lipid-free and HDL-associated apoA-I can activate the host complement pathways which is effective in killing the gastrointestinal pathogen, *Yersinia enterocolitica*. The C-terminal domain of apoA-I is the primary effector site responsible for this bactericidal property [77].

3.6. Inhibition of plasminogen recruitment

Many pathogens recruit human plasminogen (which is an integral part of the fibrinolytic system) in the course of their pathogenesis. This helps them in penetrating tissue barriers and facilitate invasion. Some pathogens even secrete plasminogen activators to amplify the effect. For example, streptokinase produced by GAS is a highly specific activator for plasminogen. Thus, it is believed that many infections can be inhibited and prevented considerably if recruitment and activation of host plasminogen by pathogens can be blocked. Lp(a) is believed to be a vital component of the host defense system in this context. Apo(a) present in Lp(a) shares a high degree of homology with plasminogen. Thus, it competes for the binding of plasminogen to pathogens. It reduces the amount of plasminogen immobilized on the pathogen surface and further inhibits the activation of plasminogen by activators such as streptokinase. In vitro studies have demonstrated the inhibition of streptokinase to catalyze the activation of plasminogen. Thus Lp(a) can help in preventing infections and promoting wound healing and repair of tissue injuries [29, 51, 78–81].

http://dx.doi.org/10.5772/67601

3.7. Chemical modification of lipoproteins

Infections and the associated inflammatory responses lead to oxidative stress and generation of reaction oxygen species (ROS). ROS induces chemical modifications in several lipoprotein species, most notable of which is oxidative changes in LDL [82]. Oxidized LDL (oxLDL) contributes to immune responses against invading pathogens in several ways. OxLDL upregulates scavenger receptor expression in macrophages, which facilitates their ingestion of Gram-positive and Gram-negative bacteria by phagocytosis. One of the oxidized components in oxLDL, namely oxidized 1-palmitoyl-2-arachidonyl-*sn*-glycero-3-phosphorylcholine (oxPAPC), modulates LPS-mediated signaling pathways in favor of the host. It inhibits LPS-induced adhesion of neutrophils to endothelial cells (thereby limiting LPS-induced tissue damage) and checks unregulated pro-inflammatory pathways [30, 82–86]. Besides, oxLDL has been shown to block cellular entry by several HCV strains [87] and malarial sporozoites [88].

Further, oxLDL elicits the production of natural antibodies against the membrane phospholipid, phosphorylcholine (PC). These anti-PC antibodies may target PC epitopes present in a broad spectrum of pathogens and provide protection against them. These include Gram-positive bacteria, Gram-negative bacteria, trematodes, nematodes, and even fungi [30, 89–93].

3.8. Acting in concert with acute-phase responses

The acute-phase response (APR), characterized by acute specific changes in concentration of plasma proteins, in response to noxious stimuli (such as infection) serves to protect the host from further injury. It helps in neutralizing the invading microbes, limits the extent of tissue damage, and promotes tissue repair and regeneration. In many instances, lipoproteins work with players of the APR in tandem and help in projecting antimicrobial defenses of the body.

For example, lipoprotein-binding protein (LBP) is an acute-phase protein carried on lipoproteins [36, 94]. It is associated with HDL, LDL, VLDL, and chylomicrons. LBP catalyzes the detoxification of bacterial toxins such as LPS and LTA by lipoproteins. LBP can modulate the effects of LPS by binding to the lipid A moiety of the latter. During infections, very high concentrations of LBP are attained, which helps in transferring LPS (and LTA) to lipoproteins for inactivation. LBP is also produced in the intestine and in the lungs where it is believed to play important roles in mobilizing local immune responses against bacterial LPS [36, 72, 94–96].

C-reactive protein (CRP) is another acute-phase protein that is associated with LDL and VLDL. Infection by the parasite *Schistosoma* leads to increase in serum CRP. CRP can activate platelets, which have cytotoxic effects against schistosomes. Such cytotoxic effects are exerted by activated monocytes as well. However, LDL binds to the surface of schistosomes, which masks them from activated monocytes. This is circumvented by oxidative changes in the parasite-bound LDL brought about by ROS from activated monocytes. OxLDL is endocytosed by the monocytes through scavenger receptors, which exposes the parasite to attack by monocytes and other immune cells [54, 55].

3.9. Redistribution of lipids to immune cells

During infection, there are quantitative and qualitative changes in plasma lipoproteins due to redistribution of lipids to the immune cells and areas of cellular injury. These changes are believed to potentiate the immune system and enhance healing in the host that helps to tide over the infective crisis [36]. For instance, there is an increase in triglyceride-rich VLDL particles, which provide lipid substrates to macrophages of the activated immune system. Similarly, there is a decrease in HDL levels. Since HDL is the central component of reverse cholesterol transport (RCT) pathway, such decrements in its level help in conserving cholesterol in peripheral sites. It has been found that during the acute phase of infection, there is an increase in apolipoprotein serum amyloid A (apoSAA) and concurrent decrease in apoA-I. ApoSAA redirects cholesterol away from catabolism in hepatocytes and delivers cholesterol to other cells. Cholesterol is required for new membrane synthesis in areas of cellular injury that accompany infections. Cholesterol may also be used for activation and proliferation of lymphocytes [97–101].

4. Lipoproteins as double-edged sword of the immune system

The immune system is a double-edged sword. Autoimmune diseases and hypersensitivity reactions are classic examples in this regard. The lipoproteins (as components of the immune system) have no exception. Lipoproteins may facilitate invasion and spread of infection by certain pathogens to the detriment of the host. Besides, lipoproteins are important risk factors for some other disorders. The following are certain examples:

- The obligate intracellular parasite, *Toxoplasma gondii*, is dependent on host cholesterol from extracellular LDL for growth and replication. The parasite resides in a special parasitophorous vacuole to which cholesterol is delivered by uptake of LDL through receptor-mediated endocytosis [102].
- There are tremendous requirements of various lipids for successful replication of the malaria parasite in the host. These requirements are met by the parasite by scavenging and modifying lipids from the host itself. Lipids such as phospholipids and free fatty acids (FFA) can be obtained from circulating lipoproteins or directly from the serum and used without further modification. Or else, the scavenged lipids are modified by elongation and desaturation reactions and subsequently incorporated as diacylglycerols and triacylglycerols [103–108].
- Similarly, a large number of viruses can hijack the host lipid and lipoprotein machinery to their benefit [109, 110]. It is increasingly appreciated that viruses can modulate lipid metabolism, composition, and signaling in the host to facilitate their entry [111–113], replication [109, 114, 115], and assembly [116–119].
- Fungal pathogens require ergosterol to grow and thrive in the host tissues. The supply of ergosterol is maintained by the endogenous sterol synthesis pathway present in the fungus. The azole group of antifungal drugs inhibits this fungal sterol synthesizing pathway.

http://dx.doi.org/10.5772/67601

However, the opportunistic fungal pathogen *Candida glabrata* can circumvent such ergosterol-deprived killing by utilizing host sterols instead. It can take up cholesterol from host circulating lipoproteins and use it for its survival in the presence of azole antifungals [120].

- Infusion of lipoproteins in volunteers has been documented to enhance growth of *Candida albicans* as well [14].
- Lipoproteins can undergo changes in their structure and composition during infections, which may be harmful to the host. As described earlier, oxLDL can help in protecting the host from the adverse effects of bacteria, viruses, and parasites. Though initially these effects are beneficial and hence desirable, yet prolonged presence of oxLDL may contribute to atherosclerosis. OxLDL plays a pivotal role in formation of lipid laden foam cells that trigger atherogenic changes [36, 82, 121, 122].
- Besides, PC, which is expressed in a number of pathogens and is targeted by natural antibodies elicited by oxLDL (described earlier), can paradoxically contribute to persistence and invasiveness of certain pathogens, such as *Haemophilus influenzae* [123, 124].
- The cholesterol-rich Lp(a) is notorious for its atherogenic and thrombotic effects. Although recent studies have described anti-infective processes in relation to Lp(a), it is nonetheless an established risk factor for cardiovascular disorders [1, 4, 5].

5. Conclusion and future directions

As our knowledge about the role of lipoproteins as crucial components of the immune system continues to advance, two types of implications for the future have emerged. First, there is the possibility of characterizing the lipoprotein-pathogen interactions in greater detail. This will lead to an improved understanding of the pathophysiological significance of these interactions and may help in elucidating novel anti-infective mechanisms. For instance, a very recent study has described serum lipoproteins as critical components for pulmonary innate defense against quorum-sensing-based pathogenesis by *Staphylococcus aureus* [125]. Second is the potential use of drug therapies to modulate lipoprotein-pathogen interactions with the aim of controlling infections. As discussed earlier, reconstituted HDL and apoA-I mimetic peptides have shown promise in this regard [46–48, 60]. Further, drugs targeting lipid metabolism have also been suggested. For example, plant extracts modulating lipoprotein metabolism have shown promising antimalarial properties [126]. Similarly, there is potential for developing therapeutics targeting fatty acid synthesis (which is required by many viruses) as broad-spectrum antiviral agents [110, 118].

To conclude, lipoproteins are increasingly recognized as important players of the host immune system. They offer a multitude of strategies to ward off infections and limit their detrimental effects in the body of the host. At times, many of these strategies act together in a complementary manner, rather than being mutually exclusive. On the other hand, an anti-infective mechanism resulting from a particular lipoprotein-pathogen interaction that may be beneficial for one specific infection may not be applicable sometimes in another infection [127]. Instead, such an interaction may promote infection and lead to untoward effects (as the previous examples show). As seen from the examples in the text, the interactions between host lipoproteins and invading pathogens are complex and multifaceted. This warrants further studies and very detailed knowledge of the different lipoprotein-pathogen interactions to design effective therapeutic options.

Conflict of interest

None.

Author details

Kaustubh Bora[1]* and Probodh Borah[2]

*Address all correspondence to: kaustubhbora1@gmail.com

1 Regional Medical Research Centre, N.E. Region (ICMR), Dibrugarh, Assam, India

2 Department of Animal Biotechnology, Bioinformatics Infrastructure Facility (BIF) and State Biotech Hub (SBT-Hub), College of Veterinary Sciences, Guwahati, Assam, India

References

[1] Voet D, Voet JG. Lipid and membranes. In: Biochemistry. 4th ed. Asia: John Wiley & Sons (Asia) Pvt Ltd; 2011. pp. 386–466.

[2] Sullivan LM, Vasan RS, Benjamin EJ, D'Agostino RB Sr. The burden of increasing worldwide cardiovascular disease. In: Fuster V, o'Rourke RA, Poole-Wilson P, Walsh RA, et al. editors. Hurst's the heart. 12th ed. New York: Tata McGraw Hill Medical; 2007.

[3] Truswell AS. Experimental pathology in St. Petersberg. In: Cholesterol and beyond the research on diet and coronary heart disease 1900–2000. 1st ed. Dordrecht, The Netherlands: Springer; 2010. pp. 5–8.

[4] Glew RH. Lipid metabolism II: pathways of metabolism of special lipids. In: Devlin TM, editor. Textbook of biochemistry with clinical correlation. 7th ed. The United States of America: John Wiley & Sons, Inc; 2011. pp. 707–749.

[5] Remaley AT, Rifai N, Warnick GR. Lipids, lipoproteins, apolipoproteins and cardiovascular risk factors. In: Burtis CA, Ashwood ER, Bruns DE, editors. Tietz textbook of clinical chemistry and molecular diagnostics. 5th ed. 1st Indian Reprint. India: Elsevier (Reed Elsevier India Private Limited); 2012. pp. 731–805.

[6] Wei Q, Wang H, Tian Y, Xu F, Chen X, Wang K. Reduced serum levels of triglyceride, very low density lipoprotein cholesterol and apolipoprotein B in Parkinson's disease patients. PloS One. 2013;**8**:e75743. doi:10.1371/journal.pone.0075743

[7] van Reedt-Dortland AKB, Giltay EJ, van Veen T, van Pelt J, Zitman FG, Penninx BWJH. Associations between serum lipids and major depressive disorder: results from the Netherlands Study of Depression and Anxiety (NESDA). J Clin Psychiatry. 2010;**71**:729–736.

[8] Marrer E, Wagner A, Montaye M, Luc G, Amouyel P, Dallongeville J, Ducimetiere P, Bingham A, Arveiler D, Velten M. Lipoprotein(a) plasma levels and the risk of cancer: the PRIME study. Eur J Cancer Prev. 2013;**22**:286–293.

[9] dos Santos CR, Domingues G, Matias I, Matos J, Fonseca I, de Almeida JM, Dias S. LDL-cholesterol signalling induces breast cancer proliferation and invasion. Lipids Health Dis. 2014;**13**:16. doi:10.1186/1476-511X-13-16

[10] Nishishita S. Studies on the fluctuation of the lipoid content of the blood in the fever period. Jpn J Exp Med. 1941;**19**:97–107.

[11] Man EB, Kartin BL, Durlacher SH, et al. The lipids of serum and liver in patients with hepatic diseases. J Clin Invest. 1945;**24**:623–643.

[12] Banerjee S, Bhaduri JN. Serum protein bound carbohydrates and lipids in cholera. Proc Soc Exp Biol Med. 1959;**101**:340–341.

[13] Hajduk SL, Hager KM, Esko JD. Human high density lipoprotein killing of African trypanosomes. Annu Rev Microbiol. 1994;**48**:139–162.

[14] Netea MG, Curfs JH, Demacker PN, Meis JF, Van Der Meer JW, Kullberg BJ. Infusion of lipoproteins into volunteers enhances the growth of *Candida albicans*. Clin Infect Dis. 1999;**28**:1148–1151.

[15] Samanovic M, Molina-Portela MP, Chessler ADC, Burleigh BA, Raper J. Trypanosome lytic factor, an antimicrobial high-density lipoprotein, ameliorates Leishmania infection. Plos Pathog. 2009;**5**:e1000276. doi:10.1371.journal.ppat.1000276

[16] Rowell J, Thompson AJ, Guyton JR, Lao XQ, McHutchison JG, McCarthy JJ, Patel K. Serum apolipoprotein C-III is independently associated with chronic hepatitis C infection and advanced fibrosis. Hepatol Int. 2012;**6**:475–481.

[17] Surewaard BGJ, Nijland R, Spaan AN, Kruijtzer JAW, de Haas CJC, van Strijp JAG. Inactivation of staphylococcal phenol soluble modulins by serum lipoprotein particles. PloS Pathog. 2012;**8**:e1002606. doi:10.1371/journal.ppat.1002606

[18] Tao J, Kang KD, Hall SD, Laube AH, Liu J, Renfrow MB, Novak J, Luo G. The serum very-low density lipoprotein serves as a restriction factor against hepatitis C virus infection. J Virol. 2015;**89**:6782–6791.

[19] Zhou L, Liu L, Yang J, Li Y, Bai W, Liu N, Li W, Gao Y, Xu L, Liu Z, Han R. LDL acts as an opsonin enhancing the phagocytosis of group A streptococcus by monocyte and whole human blood. Mol Microbiol Immunol. 2016;**205**:155–162.

[20] Sprong T, Netea MG, van Der Ley P, Verver-Jansen TJG, Jacobs LEH, Stalenhoef A, van der Meer JWM, van Deuren M. Human lipoproteins have divergent neutralizing effects on *E. coli* LPS, *N. Menigitidis* LPS, and complete Gram-negative bacteria. J Lipid Res. 2004;**45**:742–749.

[21] Wang SH, Yuan SG, Peng DQ, Zhao SP. HDL and apoA-I inhibit antigen-presentation mediated T cell activation by disrupting lipid rafts in antigen presenting cells. Atherosclerosis. 2012;**225**:105–114.

[22] Gallin JI, Kaye D, O'Leary WM. Serum lipids in infection. N Engl J Med. 1969;**281**:1081–1086.

[23] Alvarez C, Ramos A. Lipids, lipoproteins and apoproteins in serum during infection. Clin Chem. 1986;**32**:142–145.

[24] Visser BJ, Wieten RW, Nagel IM, Grobusch MP. Serum lipids and lipoproteins in malaria. Malaria J. 2013;**12**:442.

[25] Friis-Moller N, Sabin CA, Weber R, et al. Combination anti-retro viral therapy and the risk of myocardial infarction. N Engl J Med. 2003;**349**:1993–2003.

[26] Bernal E, Masia M, Padilla S, Ramos JM, Martin-Hidalgo A, Gutierrez F. Insulin resistance in HIV-infected patients receiving long term therapy with efavirenz, lopinavir/ritonavir and atazanavir. Med Clin (Barc). 2007;**129**:252–254.

[27] Bernal E, Masia M, Padilla S, Gutierrez F. High-density lipoprotein cholesterol in HIV-infected patients: evidence for an association with HIV-1 viral load, anti-retroviral therapy status, and regimen composition. AIDS Patient Care STDS. 2008;**22**: 569–575.

[28] Anastos K, Lu D, Shi Q, Tien PC, Kaplan RC, Hessol NA, Cole S, Vigen C, Cohen M, Young M, Justman J. Association of serum lipid levels with HIV serostatus, specific antiretroviral agents, and treatment regimens. J Acquir Immune Defic Syndr; 2007;**45**:34–42.

[29] Han R. Plasma lipoproteins are important components of the host defense system? J Immunol. 2009;**182**:134–174.

[30] Han R. Plasma lipoproteins are important components of the immune system. Microbiol Immunol. 2010;**54**:246–253.

[31] Ravnskov U. High cholesterol may protect against infections and atherosclerosis. QJM. 2003;**96**:927–934.

[32] Vasunilashorn S, Crimmins EM, Kim JK, Winking J, Gurven M, Kaplan H, Finch CE. Blood lipids, infection, and inflammatory markers in the Tsimane of Bolivia. Am J Hum Biol. 2010;**22**:731–740.

[33] Wiedermann U, Stemberger H, Unfried E, Widhalm K, Kundi M, Altenriederer M, Savedra M, Wiedermann G. Intestinal worm burden and serum cholesterol or lipid concentration in a Shipibo population (Peru). Zentralbl Bakteriol. 1991;**275**:279–286.

[34] Femling JK, West SD, Hauswald EK, Gresham HD, Hall PR. Nosocomial infections after severe trauma are associated with lower apolipoproteins B and AII. J Trauma Acute Care Surg. 2013;**74**:1067–1073.

[35] Singh IP, Chopra AK, Coppenhaver DH, Ananatharamaiah GM, Baron S. Lipoproteins account for part of the broad non-specific antiviral activity of human serum. Antiviral Res. 1999;**42**:211–218.

[36] Khovidhunkit W, Kim MS, Memon RA, Shigenaga JK, Moser AH, Feingold KR, Grunfeld C. Effects of infection and inflammation on lipid and lipoprotein metabolism: mechanisms and consequences to the host. J Lipid Res. 2004;**45**:1169–1196.

[37] Shortridge KF, Ho WK, Oya A, Kobayashi M. Studies on the inhibitory activities of human serum lipoproteins for Japanese encephalitis virus. Southeast Asian J Trop Med Public Health. 1975;**6**:461–466.

[38] Shortridge KF, Ho WK. Comparison of the activities in inhibition of haemagglutination by different togaviruses for human serum lipoproteins and their constituents. J Gen Virol. 1976;**33**:523–527.

[39] Chisari FV, Curtiss LK, Jensen FC. Physiologic concentrations of normal human plasma lipoproteins inhibit the immortalization of peripheral B lymphocytes by the Epstein-Barr virus. J Clin Invest. 1981;**68**:329–336.

[40] Seganti L, Grassi M, Mastromarino P, Pana A, Superti F, Orsi N. Activity of human serum lipoproteins on the infectivity of rhabdoviruses. Microbiologica. 1983;**6**:91–99.

[41] Huemer HP, Menzel HJ, Potratz D, Brake B, Falke D, Utermann G, Dierich MP. *Herpes simplex* virus binds to human serum lipoprotein. Intervirology. 1988; **29**:68–76.

[42] Emancipator K, Csako G, Elin RJ. *In-vitro* inactivation of bacterial endotoxin by human lipoproteins and apolipoproteins. Infect Immun. 1992;**60**:596–601.

[43] Ulevitch RJ, Johnston AR, Weinstein DB. New function for high density lipoproteins. Their participation in intravascular reactions of bacterial lipopolysaccharides. J Clin Invest. 1979; **64**:1516–1524.

[44] Eichbaum EB, Harris HW, Kane JP, Rapp JH. Chylomicrons can inhibit endotoxin activity *in-vitro*. J Surg Res. 1991;**51**:413–416.

[45] Harris HW, Eichbaum EB, Kane JP, Rapp JH. Detection of endotoxin in triglyceride-rich lipoproteins *in-vitro*. J Lab Clin Med. 1991;**118**:186–193.

[46] Hubsch AP, Casas AT, Doran JE. Protective effects of reconstituted high-density lipoprotein in rabbit Gram-negative bacteremia models. J Lab Clin Med. 1995;**126**:548–558.

[47] Parker TS, Levine DM, Chang JC, Laxer J, Coffin CC, Rubin AL. Reconstituted high-density lipoprotein neutralizes gram-negative bacterial lipopolysaccharides in human whole blood. Infect Immun. 1995;**63**:253–258.

[48] Casas AT, Hubsch AP, Doran JE. Effects of reconstituted high-density lipoprotein in persistent gram-negative bacteremia. Am Surg. 1996;**62**:350–355.

[49] Bhakdi S, Tranum-Jensen J, Utermann G, Fussle R. Binding and partial inactivation of *Staphylococcus aureus* alphatoxin by human plasma low density lipoprotein. J Biol Chem. 1983;**258**:5899–5904.

[50] Makoveichuk E, Cherepanov P, Lundberg S, Forsberg A, Olivecrona G. pH6 antigen of *Yersinia pestis* interacts with plasma lipoproteins and cell membranes. J Lipid Res. 2003;**44**:320–330.

[51] Han R, Caswell CC, Lukomska E, Keene DR, Pawlowski M, Bujnicki JM, Kim JK, Lukomski S. Binding of the low-density lipoprotein by streptococcal collagen-like protein Scl1 of *Streptococcus pyogenes*. Mol Microbiol. 2006;**61**:351–367.

[52] Roselaar SE, Daugherty A. Apolipoprotein E-deficient mice have impaired innate immune responses to *Listeria monocytogenes in-vivo*. J Lipid Res. 1998;**39**:1740–1743.

[53] Martens GW, Arikan MC, Lee J, Ren F, Vallerskog T, Kornfeld H. Hypercholesterolemia impairs immunity to tuberculosis. Infect Immun. 2008;**76**:3464–3472.

[54] Bout D, Joseph M, Pontet M, Vorng H, Deslee D, Capron A. Rat resistance to schistosomiasis: platelet-mediated cytotoxicity induced by C-reactive protein. Science. 1986;**231**:153–156.

[55] Xu X, Remold HG, Caulfield JP. Potential role for scavenger receptors of human monocytes in the killing of *Schistosoma mansoni*. Am J Pathol. 1993;**142**:685–689.

[56] Sinnis P, Willnow TE, Briones MR, Herz J, Nussenzweig V. Remnant lipoproteins inhibit malaria sporozoite invasion of hepatocytes. J Exp Med. 1996;**184**:945–954.

[57] Mastromarino P, Seganti L, Orsi N. Relationship between enzymatic modifications of serum low density lipoproteins and their haemagglutination inhibiting activity towards Sindbis virus. Arch Virol. 1980;**65**:37–44.

[58] Martin I, Dubois MC, Saermark T, Ruysschaert JM. Apolipoprotein A-1 interacts with the N-terminal fusogenic domains of SIV (simian immunodeficiency virus) GP32 and HIV (human immunodeficiency virus) GP41: implications in viral entry. Biochem Biophys Res Commun. 1992;**186**:95–101.

[59] Srinivas RV, Birkedal B, Owens RJ, Anantharamaiah GM, Segrest JP, Compans RW. Antiviral effects of apolipoprotein A-I and its synthetic amphipathic peptide analogs. Virology. 1990;**176**:48–57.

[60] Owens BJ, Anantharamaiah GM, Kahlon JB, Srinivas RV, Compans RW, Segrest JP. Apolipoprotein A-I and its amphipathic helix peptide analogues inhibit human immunodeficiency virus-induced syncytium formation. J Clin Invest. 1990;**86**:1142–1150.

[61] Fischer DG, Tal N, Novick D, Barak S, Rubinstein M. An antiviral soluble form of the LDL receptor induced by interferon. Science. 1993;**262**:250–253.

[62] Hofer F, Gruenberger M, Kowalski H, Machat H, Huettinger M, Kuechler E, Blass D. Members of the low density lipoprotein receptor family mediate cell entry of a minor group common cold virus. Proc Natl Acad Sci USA. 1994;**91**:1839–1842.

[63] Marlovits TC, Abrahamsberg C, Blaas D. Very-low density lipoprotein receptor fragment shed from HeLa cells inhibits human rhinovirus infection. J Virol. 1998;**72**:10246–10250.

[64] Netea MG, Joosten LAB, Keuter M, Wagener F, Stalenhoef AFH, van der Meer JWM, Kullberg BJ. Circulating lipoproteins are a crucial component of host defense against invasive *Salmonella typhimurium* infection. PLoS One. 2009;**4**:e4237. doi:10.1371/journal.pone.0004237

[65] Netea MG, de Bont N, Demacker PN, Kullberg BJ, Jacobs LE, Verver-Jansen TJ, Stalenhoef AF, Van der Meer JW. Lipoprotein(a) inhibits lipopolysaccharide-induced tumor necrosis factor alpha production by human mononuclear cells. Infect Immun. 1998;**66**:2365–2367.

[66] Harris HW, Grunfeld C, Feingold KR, Rapp JH. Human very low density lipoproteins and chylomicrons can protect against endotoxin-induced death in mice. J Clin Invest. 1990;**86**:696–702.

[67] Victorov AV, Medvedeva NV, Gladkaya EM, Morozkin AD, Podrez EA, Kosykh VA, Yurkiv VA. Composition and structure of lipopolysaccharide-human plasma low density lipoprotein complex. Analytical ultracentrifugation, 31P-NMR, ESR and fluorescence spectroscopy studies. Biochim Biophys Acta. 1989;**984**:119–127.

[68] Harris HW, Grunfeld C, Feingold KR, Read TE, Kane JP, Jones AL, Eichbaum EB, Bland GF, Rapp JH. Chylomicrons alter the fate of endotoxin, decreasing tumor necrosis factor release and preventing death. J Clin Inves. 1993;**91**:1028–1034.

[69] Read TE, Harris HW, Grunfeld C, Feingold KR, Calhoun MC, Kane JP, Rapp JH. Chylomicrons enhance endotoxin excretion in bile. Infect Immun. 1993;**61**:3496–3502.

[70] Read TE, Grunfeld C, Kumwenda Z, Calhoun MC, Kane JP, Feingold KR, Rapp JH. Triglyceride-rich lipoproteins improve survival when given after endotoxin in rats. Surgery. 1995;**117**:62–67.

[71] Kitchens RL, Wolfbauer G, Albers JJ, Munford RS. Plasma lipoproteins promote the release of bacterial lipopolysaccharides from the monocyte cell surface. J Biol Chem. 1999;**274**:34116–34122.

[72] Grunfeld C, Marshall M, Shigenaga JK, Moser AH, Tobias P, Feingold KR. Lipoproteins inhibit macrophage activation by lipoteichoic acid. J Lipid Res. 1999;**40**:245–252.

[73] Raper J, Fung R, Ghiso J, Nussenzweig V, Tomlinson S. Characterization of a novel trypanosome lytic factor from human serum. Infect Immun. 1999;**67**:1910–1916.

[74] Raper J, Portela MP, Lugli E, Frevert U, Tomlinson S. Trypanosome lytic factors: novel mediators of human innate immunity. Curr Opin Microbiol. 2001;**4**:402–408.

[75] Vanhamme L, Paturiaux-Hanocq F, Poelvoorde P, Nolan DP, Lins L, Van Den Abbeele J, Pays A, Tebabi P, Van Xong H, Jacquet A, Moguilevsky N, Dieu M, Kane JP, De Baetselier P, Brasseur R, Pays E. Apolipoprotein L-I is the trypanosome lytic factor of human serum. Nature. 2003;**422**:83–87.

[76] Miao Q, Santamaria C, Bailey D, Genest J, Ward BJ, Ndao M. Apolipoprotein A-I truncations in Chagas disease are caused by cruzipain, the major cysteine protease of *Trypanosoma cruzi*. Am J Pathol. 2014;**184**:876–984.

[77] Biedzka-Sarek M, Metso J, Kateifides A, et al. Apolipoprotein A-I exerts bactericidal activity against *Yersinia enterocolitica* serotype O:3. J Biol Chem. 2011;**286**:38211–38219.

[78] Lahteenmaki K, Edelman S, Korhonen TK. Bacterial metastasis: the host plasminogen system in bacterial invasion. Trends Microbiol. 2005;**13**:79–85.

[79] Edelberg JM, Gonzalez-Gronow M, Pizzo SV. 1989. Lipoprotein a inhibits streptokinase-mediated activation of human plasminogen. Biochemistry. 1989;**28**:2370–2374.

[80] Sun H, Ringdahl U, Homeister JW, Fay WP, Engleberg NC, Yang AY, Rozek LS, Wang X, Sjobring U, Ginsburg D. Plasminogen is a critical host pathogenicity factor for group A streptococcal infection. Science; 2004;**305**:1283–1286.

[81] Sun H. The interaction between pathogens and the host coagulation system. Physiology (Bethesda). 2006;**21**:281–288.

[82] Memon RA, Staprans I, Noor M, Holleran WM, Uchida Y, Moser AH, Feingold KR, Grunfeld C. 2000. Infection and inflammation induce LDL oxidation *in-vivo*. Arterioscler Thromb Vasc Biol. 2000;**20**:1536–1542.

[83] Miller YI, Viriyakosol S, Binder CJ, Feramisco JR, Kirkland TN, Witztum JL. Minimally modified LDL binds to CD14, induces macrophage spreading via TLR4/MD-2, and inhibits phagocytosis of apoptotic cells. J Biol Chem. 2003;**278**:1561–1568.

[84] Bochkov VN, Kadl A, Huber J, Gruber F, Binder BR, Leitinger N. Protective role of phospholipid oxidation products in endotoxin-induced tissue damage. Nature. 2002;**419**:77–81.

[85] Bengtsson T, Karlsson H, Gunnarsson P, Skoglund C, Elison C, Leanderson P, Lindahl M. The periodontal pathogen *Porphyromonas gingivalis* cleaves apoB-100 and increases the expression of apoM in LDL in whole blood leading to cell proliferation. J Intern Med. 2008;**263**:558–571.

[86] Walton KA, Cole AL, Yeh M, Subbanagounder G, Krutzik SR, Modlin RL, Lucas RM, Nakai J, Smart EJ, Vora DK, Berliner JA. Specific phospholipid oxidation products inhibit ligand activation of Toll-like receptors 4 and 2. Arterioscler Thromb Vasc Biol. 2003;**23**:1197–1203.

[87] Von Hahn T, Lindenbach BD, Boullier A, Quehenberger O, Paulson M, Rice CM, Mckeating JA. Oxidized low-density lipoprotein inhibits hepatitis C virus cell entry in human hepatoma cells. Hepatology. 2006;**43**:932–942.

[88] Rodrigues CD, Hannus M, Prudencio M, Martin C, Goncalves LA, Portugal S, Epiphanio S, Akinc A, Hadwiger P, Jahn-Hofmann K, Rohl I, Van Gemert GJ, Franetich JF, Luty AJ, Sauerwein R, Mazier D, Koteliansky V, Vornlocher HP, Echeverri CJ, Mota MM. Host scavenger receptor SR-BI plays a dual role in the establishment of malaria parasite liver infection. Cell Host Microbe. 2008;**4**:271–282.

[89] Gillespie SH, Mcwhinney PH, Patel S, Raynes JG, Mcadam KP, Whiley RA, Hardie JM. Species of alpha-hemolytic streptococci possessing a C-polysaccharide phosphorylcholine-containing antigen. Infect Immun. 1993;**61**:3076–3077.

[90] Weiser JN, Goldberg JB, Pan N, Wilson L, Virji M. The phosphorylcholine epitope undergoes phase variation on a 43-kilodalton protein in *Pseudomonas aeruginosa* and on pili of *Neisseria meningitidis* and *Neisseria gonorrhoeae*. Infect Immun. 1998;**66**:4263–4267.

[91] Chou MY, Hartvigsen K, Hansen LF, Fogelstrand L, Shaw PX, Boullier A, Binder CJ, Witztum JL. Oxidation-specific epitopes are important targets of innate immunity. J Intern Med. 2008;**263**:479–488.

[92] Kolberg J, Hoiby EA, Jantzen E. Detection of thephosphorylcholine epitope in streptococci, Haemophilus and pathogenic Neisseriae by immunoblotting. Microb Pathog. 1997;**22**:321–329.

[93] Serino L, Virji M. Genetic and functional analysis of the phosphorylcholine moiety of commensal Neisseria lipopolysaccharide. Mol Microbiol. 2002;**43**:437–448.

[94] Park CT, Wright SD. Plasma lipopolysaccharides binding protein is found associated with a particle containing apolipoprotein A-I, phospholipid, and factor H-related proteins. J Biol Chem. 1996;**271**:18054–18060.

[95] Vreugdenhil AC, Snoek AM, van't Veer C, Greve JW, Buurman WA. LPS-binding protein circulates in association with apoB-containing lipoproteins and enhances endotoxin-LDL/VLDL interaction. J Clin Invest. 2001;**107**:225–234.

[96] Vreugdenhil AC, Rousseau CH, Hartung T, Greve JW, van't Veer C, Buurman WA. Lipopolysaccharide (LPS)-binding protein mediates LPS detoxification by chylomicrons. J Immunol. 2003;**170**:1399–1405.

[97] Cuthbert JA, Lipsky PE. Regulation of lymphocyte proliferation by cholesterol: the role of endogenous sterol metabolism and low density lipoprotein receptors. Int J Tissue React. 1987;**9**:447–457.

[98] Kisilevsky R, Subrahmanyan L. Serum amyloid A changes high density lipoprotein's cellular affinity. A clue to serum amyloid A's principal function. Lab Invest. 1992;**66**:778–785.

[99] Oiknine J, Aviram M. Increased susceptibility to activation and increased uptake of low density lipoprotein by cholesterol-loaded macrophages. Arterioscler Thromb; 1992;**12**:745–753.

[100] Hauton D, Evans RD. Utilisation of fatty acid and triacylglycerol by rat macrophages: the effect of endotoxin. Cell Physiol Biochem. 2002;**12**:293–304.

[101] Funk JL, Feingold KR, Moser AH, Grunfeld C. Lipopolysaccharide stimulation of RAW 264.7 macrophages induces lipid accumulation and foam cell formation. Atherosclerosis. 1993;**98**:67–82.

[102] Coppens I, Sinai AP, Joiner KA. *Toxoplasma gondii* exploits host low density lipoprotein receptor-mediated endocytosis for cholesterol acquisition. J Cell Biol. 2000;**149**:167–180.

[103] Palacpac NMQ, Hiramine Y, Mi-Ichi F, et al. Developmental-stage-specific triacylglycerol biosynthesis, degradation and trafficking as lipid bodies in *Plasmodium falciparum*-infected erythrocytes. J Cell Science. 2004;**117**:1469–1480.

[104] Mi-Ichi F, Kita K, Mitamura T. Intraerythrocytic *Plasmodium falciparum* utilize a broad range of serum-derived fatty acids with limited modifications for their growth. Parasitology. 2006;**133**:399–410.

[105] Moll GN, Vial HJ, Ancelin L, et al. Phospholipid uptake by *Plasmodium knowlesi* infected erythrocytes. FEBS Lett. 1988;**232**:341–346.

[106] Krishnegowda G, Gowda DC. Intraerythrocytic *Plasmodium falciparum* incorporates extraneous fatty acids to its lipids without any structural modification. Mol Biochem Parasitol. 2003;**132**:55–58.

[107] Grellier P, Rigomier D, Clavey V, Fruchart JC, Schrevel J. Lipid traffic between high density lipoproteins and *Plasmodium falciparum*-infected red blood cells. J Cell Biol. 1991;**112**:267–277.

[108] Vial HJ, Thuet MJ, Broussal JL, Philippot JR. Phospholipid biosynthesis by *Plasmodium knowlesi*-infected erythrocytes: the incorporation of phospohlipid precursors and the identification of previously undetected metabolic pathways. J Parasitol. 1982;**68**:379–391.

[109] Heaton NS, Perera R, Berger KL, Khadka S, Lacount DJ, Kuhn RJ, Randall G. Dengue virus non-structural protein 3 redistributes fatty acid synthase to sites of viral replication and increases cellular fatty acid synthesis. Proc Natl Acad Sci USA. 2010;**107**:17345–17350.

[110] Heaton NS, Randall G. Dengue virus and autophagy. Viruses. 2011;**3**:1332–1341.

[111] Agnello V, Abel G, Elfahal M, Knight GB, Zhang QX. Hepatitis C virus and other flaviviridae viruses enter cells via low density lipoprotein receptor. Proc Natl Acad Sci USA. 1999;**96**:12766–12771.

[112] Von Hahn T, Rice CM. Hepatitis C virus entry. J Biol Chem. 2008;**283**:3689–3693.

[113] Sabahi A. Hepatitis C virus entry: the early steps in the viral replication cycle. Virol J. 2009;**6**:117. doi:10.1186/1743-422X-6-117

[114] Mackenzie JM, Khromykh AA, Parton RG. Cholesterol manipulation by West Nile virus perturbs the cellular immune response. Cell Host Microb. 2007;**2**:229–239.

[115] Rothwell C, Lebreton A, Young Ng C, Lim JY, Liu W, Vasudevan S, Labow M, Gu F, Gaither LA. Cholesterol biosynthesis regulation modulates dengue viral replication. Virology. 2009;**389**:8–19.

[116] Nguyen DH, Hildreth JE. Evidence for budding of human immunodeficiency virus type 1 selectively from glycolipid-enriched membrane lipid rafts. J Virol. 2000;**74**:3264–3272.

[117] Zheng YH, Plemenitas A, Fielding CJ, Peterlin BM. Nef increases the synthesis of and transports cholesterol to lipid rafts and HIV-1 progeny viruses. Proc Natl Acad Sci USA. 2003;**100**:8460–8465.

[118] Munger J, Bennett BD, Parikh A, Feng XJ, McArdle J, Rabitz HA, Rabinowitz JD. Systems level metabolic flux profiling identifies the fatty acid synthesis as a target for anti-viral therapy. Nat Biotechnol. 2006;**26**:1179–1186.

[119] Samsa MM, Mondotte JA, Iglesias NG, Assuncao-Miranda I, Barbosa-Lima G, Da Poian AT, Bozza PT, Gamarnik AV. Dengue virus capsid protein usurps lipid droplets for viral particle formation. PLOS Pathog. 2009;**5**:e1000632. doi:10.1371/journal.ppat.1000632

[120] Nagi M, Tanabe K, Nakayama H, Yamagoe S, Umeyama T, Oura T, Ohno H, Kajiwara S, Miyazaki Y. Serum cholesterol promotes the growth of *Candida glabrata*. J Infect Chemother. 2013;**19**:138–143.

[121] Ishigaki Y, Katagiri H, Gao J, Yamada T, Imai J, Uno K, Hasegawa Y, Kaneko K, Ogihara T, Ishihara H, Sato Y, Takikawa K, Nishimichi N, Matsuda H, Sawamura T, Oka Y. Impact of plasma oxidized low-density lipoprotein removal on atherosclerosis. Circulation. 2008;**118**:75–83.

[122] Shoenfeld Y, Wu R, Dearing LD, Matsuura E. Are anti-oxidized low-density lipoprotein antibodies pathogenic or protective? Circulation. 2004;**110**:2552–2558.

[123] Weiser JN, Pan N, Mcgowan KL, Musher D, Martin A, Richards J. Phosphorylcholine on the lipopolysaccharide of *Haemophilus influenzae* contributes to persistence in the respiratory tract and sensitivity to serum killing mediated by C-reactive protein. J Exp Med. 1998;**187**:631–640.

[124] Lysenko E, Richards JC, Cox AD, Stewart A, Martin A, Kapoor M, Weiser JN. The position of phosphorylcholine on the lipopolysaccharide of *Haemophilus influenzae* affects binding and sensitivity to C-reactive protein-mediated killing. Mol Microbiol. 2000;**35**:234–245.

[125] Manifold-Wheeler BC, Elmore BO, Triplett KD, Castleman MJ, Otto M, Hall PR. Serum lipoproteins are critical for pulmonary innate defense against *Staphylococcus aureus* quorum sensing. J Immunol. 2016;**196**:328–335.

[126] Oluba OM, Olusola AO, Eidanghe GO, Babatola LJ, Onyeneke EC. Modulation of lipoprotein cholesterol levels in *Plasmodium berghei* malarial infection by crude aqueous extract of Ganoderma lucidum. Cholesterol. 2012;**2012**:536396. doi:10.1155/2012/536396

[127] Feingold KR, Grunfeld C. Lipids: a key player in the battle between the host and microorganisms. J Lipid Res. 2012;**53**:2487–2489.

Section 4

Lipoproteins in Specific Diseases

Chapter 6

Lipoproteins and Diseases of the Brain

Kwang-Min Kim and G. Tayhas R. Palmore

Additional information is available at the end of the chapter

http://dx.doi.org/10.5772/67053

Abstract

Apolipoprotein E4 (apoE4) and outer surface protein A (ospA) are pathogenic lipoproteins involved in the progression of Alzheimer's disease and Lyme neuroborreliosis, respectively. Results from previous studies indicate that apoE4 exhibits neurotoxicity by activating amyloid beta pathways, and ospA causes damage to the brain by stimulating immune activity of microglia and astrocytes. These results, however, lack information about the specific interactions that develop between neurons and these two lipoproteins. It is essential to investigate the effect of these lipoproteins on neuronal morphology and function to better understand the mechanism of damage and disease of the brain. This chapter summarizes previous studies on the role of apoE4 and ospA in diseases of the brain and discusses experimental results from our own work that suggests new roles for apoE4 and ospA in neuronal outgrowth and synaptic loss.

Keywords: apolipoprotein E4, bacterial outer surface protein A, neurodegeneration, neuroinflammation, nerve regeneration, synaptic loss

1. Introduction

Lipoproteins in the brain are involved in the onset and progression of neurodegenerative diseases (e.g., Alzheimer's disease) [1, 2] and neuroinflammatory disorders (e.g., neuroborreliosis) [3, 4]. These lipoproteins are either endogenously expressed by astrocytes [5] and microglia [6, 7] or exogenously produced by bacterial pathogens (e.g., *Borrelia burgdorferi, Streptococcus pneumoniae*) [8].

The most abundant endogenous lipoproteins in the brain include apolipoprotein E (apoE) and apoJ [2]. These endogenous lipoproteins mediate transport of lipids between various cells in the brain to maintain and regulate the brain structure and function [9, 10]. The apoE isoform, apoE4, has been investigated intensively because previous studies showed that lipidation of apoE4 (i.e., apoE4 carrying cholesterol and phospholipids) is the major risk factor indicative of the onset of Alzheimer's disease (AD) [11].

The exogenous lipoprotein most studied in the brain is the bacterial outer surface protein A (ospA), which is produced by *B. burgdorferi* [12, 13]. *B. burgdorferi* causes Lyme disease, which is the most common tick-borne infection in Europe and in North America [14]. A recent study using rats infected with *B. burgdorferi* demonstrated that *B. burgdorferi* was observed across the blood-brain barrier (BBB) and that the expression level of ospA was augmented significantly in the brain [4].

Thus, apoE and ospA have been of interest to both scientists and clinicians who seek to develop new strategies for treatment of brain injuries and brain disorders induced by these pathogenic lipoproteins. It still remains unclear however, if apoE and ospA interact directly with neurons to disrupt the structure and function of the brain, whereas it is documented extensively that these lipoproteins induce pathological states via amyloid beta (Aβ) aggregation [15, 16] and immune activation of microglia and astrocytes [17, 18]. To address the absence of direct evidence of interaction between lipoproteins and neurons, we have studied the effect of apoE4 and ospA on neurons in terms of axonal outgrowth and synaptic loss. This chapter discusses these findings and the potential new roles of apoE4 and ospA in the context of previous studies on these lipoproteins in neurodegeneration and neuroinflammation.

2. ApoE4 and neuronal outgrowth

2.1. Lipidation of apoE isoforms

ApoE transports and clears lipids from one cell to another to maintain lipid homeostasis of the brain [9, 10]. To carry lipids (e.g., cholesterol, phospholipids, and lipoproteins), apoE is lipidated (i.e., lipid-bound apoE) by adenosine triphosphate (ATP)-binding cassette A1 (ABCA1) transporters on astrocytes [19] (**Figure 1a**). The lipidation status of apoE depends on its three isoforms (i.e., apoE2, apoE3, and apoE4) coded by three alleles (ε2, ε3, and ε4 of *APOE* gene) on chromosome 19 [20]. The three isoforms of apoE differ from one another by amino acid interchanges at two residue sites (**Table 1**).

These minor variations cause a change in the structure and function of apoE, which eventually leads to distinct disease mechanisms in AD [21]. ApoE4 has an arginine at residue 112 that connects the N terminus (Arg 61) to the C terminus (Glu 255) to form a folded structure of apoE called *domain interaction* [22]. ApoE2 and apoE3 have a cysteine at residue 112, which is less likely to create the folded structure of domain interaction. The presence of domain interaction results in distinct lipid-bound forms among apoE isoforms. ApoE4 binds preferentially to larger lipid particles due to its folded structure, which interferes with internalization of lipids into neurons [11]. In contrast, apoE2 and E3 bind to various sizes of lipids in more ways that are efficient and thus facilitate lipid transport between cells in the brain. The lipidated apoE can be internalized into cells in the brain (i.e., astrocytes, microglia, and neurons) through the family of low-density lipoprotein receptors (LDLR), low-density lipoprotein receptor-related protein 1 (LRP1), or heparan sulfate proteoglycans (HSPGs) [11, 16].

http://dx.doi.org/10.5772/67053

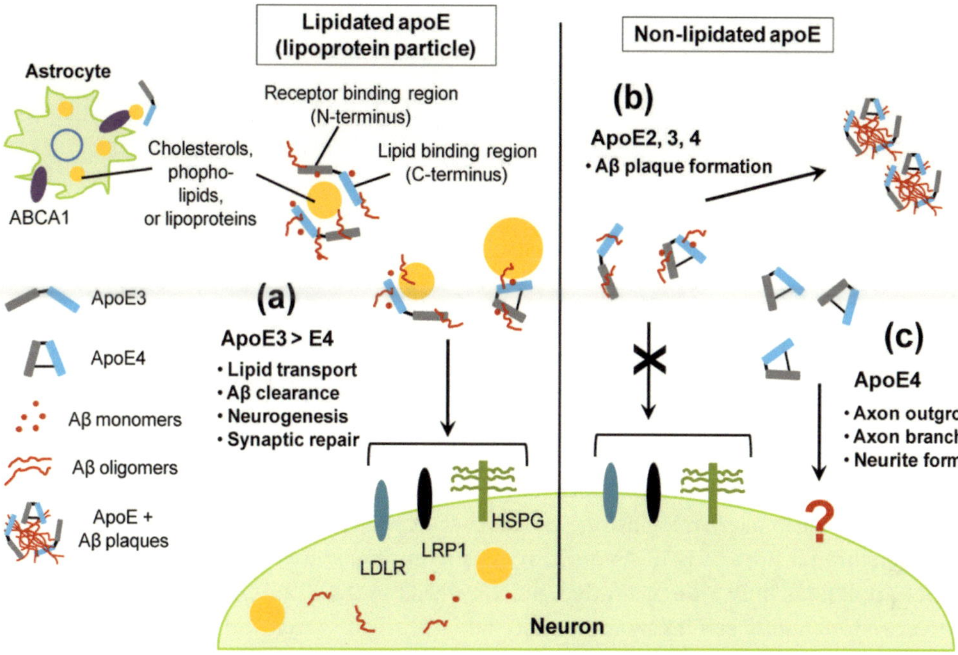

Figure 1. The interaction between apoE and neurons illustrated. (a) ApoE transports lipids to neurons by forming lipopeptide particles (i.e., lipidation of apoE). ApoE is lipidated by ATP-binding cassette A1 (ABCA1) transporters of astrocytes. The lipid-binding affinity of apoE4 is different from that of apoE2 and apoE3 because of structural differences in its *domain interaction*. Both lipid and apoE can bind to Aβ monomers and oligomers. The Aβ-lipidated apoE2/3 complex can be internalized by LDLR, LRP1, or HSPG, which clears Aβ. The efficiency of internalizing large lipid-bound apoE4 into cells is low, which increases the probability of Aβ plaque formation because of poor Aβ clearance. (b) ApoE alone can bind to Aβ monomers and oligomers regardless of its isoforms. The Aβ-nonlipidated apoE complex increases the probability of forming Aβ plaques because nonlipidated apoE cannot be internalized via LDLR or LRP1. (c) Nonlipidated apoE4 enhances neuronal adhesion, axon outgrowth, and neurite branching. The receptor in neurons that regulates growth-enhancing effects of nonlipidated apoE4 remains unknown. Abbreviations: ABCA1, ATP-binding cassette A1 transporter; LDLR, low-density lipoprotein receptor; LRP1, low-density lipoprotein receptor-related protein 1; HSPG, heparan sulfate proteoglycans.

ApoE isoforms	ApoE amino acid residue	
	112	**158**
ApoE2	Cys	Cys
ApoE3	Cys	Arg
ApoE4	Arg	Arg

Table 1. Differences of apoE isoforms in amino acid residues.

When the lipidated apoE is internalized into cells, Aβ monomers and oligomers are also cleared because they bind to both lipids and apoE at residues 12–28 [23]. Thus, Aβ clearance is dependent on the structural difference of apoE isoforms, and this mechanism helps to prevent the Aβ aggregation that is associated with the progression of AD. The Aβ-bound apoE, however, forms aggregates regardless of the isoform of apoE when they are not lipidated and thus, are not internalized [16, 23] (**Figure 1b**).

2.2. Nonlipidated apo E4 and neuronal outgrowth

When lipidated, apoE4 is known to be toxic to neurons through various pathogenic pathways such as Aβ aggregation and apoE fragment formation [21]. The effect of apoE4 on neurons when it is not lipidated, however, remains unclear. To address this knowledge gap, the effect of apoE4 on neuronal outgrowth was studied *in vitro* without lipids in the medium [24]. This study compared neuronal responses to various culture substrates including glass, laminin-coated glass, and apolipoprotein E4-coated glass by quantifying key neuronal outgrowth parameters in terms of cell adhesion, axon length, number of neurites, and number of branches on axons. The results of this study demonstrated that apoE4 not only enhances neuronal adhesion but also significantly increases axon outgrowth and branching when compared to laminin, a protein that is recognized as one of the best extracellular matrix (ECM) proteins for enhancing neuronal growth [25]. As such, results from this study contradict the prevailing view that apoE4 has only a degenerative effect on cells in the brain. Although apoE4 when lipidated predominantly exhibits neurotoxicity when studied *in vivo* and in clinical models, it should be considered that both lipidated and nonlipidated apoE in these models constantly interact with neurons to mediate brain activity. Thus, the results from this study provide a complementary mechanism of action of apoE. In addition, the neuron-growing potential of apoE4 can be applied to transplantable therapeutic systems using stem cells or microstructure devices prior to interaction of lipids *in vivo*.

The mechanism by which nonlipidated apoE mediates axon outgrowth and branching remains elusive, whereas lipidated apoE is known to interact with cells via LRP1, LDLR, or HSPG [16]. It has been reported that apoE does not bind to LDLR or LRP1 without lipidation [26, 27]. Integrin and HSPGs also were tested for their involvement in apoE4-induced axon outgrowth by inhibiting these receptors. Neither of these receptors was found to be responsible for apoE4-induced neuronal outgrowth (**Figure 1c**). The mechanism of interaction between neurons and nonlipidated apoE4 is the subject of ongoing studies.

3. Bacterial lipoprotein and synaptic loss

3.1. Bacterial lipoproteins and neuroinflammation

Bacterial surface components including lipoproteins and lipopolysaccharide (LPS) have been reported to be elevated in the cerebrospinal fluid (CSF) of patients suffering from a bacterial infection such as bacterial meningitis [28]. These components can cause neuropsychiatric manifestations such as lymphocytic meningitis, cranial and peripheral neuropathy, and cerebral infarcts [29, 30]. When compared to LPS, bacterial lipoproteins activate inflammatory pathways more vigorously [31], leading to more severe damage to tissue [32]. Bacterial lipoproteins still remain in the tissue even after the degradation of bacteria by antibiotic therapies [33, 34]. As a result, many studies suggest that minimizing the production of bacterial proteins or inhibiting bacterial protein synthesis is more effective at preventing neural injury from bacterial infections in animal models or patients [35, 36] than simply using antibiotics to kill bacteria. Bacterial lipoproteins in the brain trigger microglia activation via the toll-like receptors (TLRs) to produce inflammatory mediators (e.g., cytokines and reactive oxygen species) [37–39] and induce migration of immune cells across the BBB [40, 41]. The result is damaged brain tissue including cell death of astrocytes, oligodendrocytes, and neurons [42, 43].

The outer surface protein (osp) is the most studied bacterial lipoprotein that includes ospA, ospB, and ospC from *B. burgdorferi* [12, 44, 45]. Three palmitoyl groups (i.e., the lipid portion) at the N-terminus of the peptide is responsible for immune activation and tissue injury [46, 47], whereas the peptide portion of ospA is not effective at activating immune pathways [32]. Thus, tripalmitoyl-S-glyceryl-cysteine (Pam3-Cys), a synthetic lipopeptide mimicking the N-termini of osp, is often used for studying bacterial infection in a wide range of research fields involving immunology and neuroscience [48, 49]. Although all of ospA, ospB, and ospC share common immune pathways (e.g., NF-κB activation) via TLR2, ospA shows higher toxicity to tissues when compared with ospB and ospC [32]. The reason for distinct toxic effects among these different lipoproteins continues to remain elusive.

3.2. OspA and presynaptic loss

OspA from *B. burgdorferi* is able to cross the BBB by binding to CD40 of brain-microvascular endothelial cells [4]. OspA in the brain activates TLR2 on microglia and astrocytes, which initiates immune activity and causes damage to brain tissue [14, 50]. However, information regarding the interaction between ospA and neurons is lacking because the expression level of TLR2 in neurons is extremely low when compared with that of microglia or astrocytes. Thus, the interaction between ospA and neurons has been overlooked [51, 52]. To address this question, the effect of ospA on neurons has been investigated with a specific focus on synaptic loss. The density and transmission of synapses are considered to be the key parameters in determining the functional state of brain tissue (e.g., information processing and storage) because neurons transmit electrical and biochemical signals to adjacent neurons through the synapse. The signal-sending synapse (i.e., presynapse) is located on the axon and the signal-receiving synapse (i.e., postsynapse) is located on the dendrite of a neuron. If neurons lose one of these synapses or have misaligned synapses, the brain cannot function properly even when neurons survive from brain injuries or diseases. Thus, the change in pre- and postsynaptic density was quantified following treatment of cultures of rat E18 hippocampal neurons with ospA (2 μM) for 24 h (**Figure 2**). The quantification of synaptic density was determined by counting the number of synaptic sites (i.e., synapsin or postsynaptic density protein 95 (PSD-95)) in a randomly selected secondary dendrite. OspA expressed from *Escherichia coli* (prepared by the Biomaterials and Advanced Drug Delivery Laboratory at Stanford University) showed that ospA significantly decreased the number of presynaptic sites (i.e., synapsin) ($p = 0.04$), whereas it did not affect the number of postsynaptic sites (i.e., PSD-95) ($p > 0.05$) (**Figure 2**). This result suggests that ospA directly disrupts neuronal function by damaging presynapses exclusively.

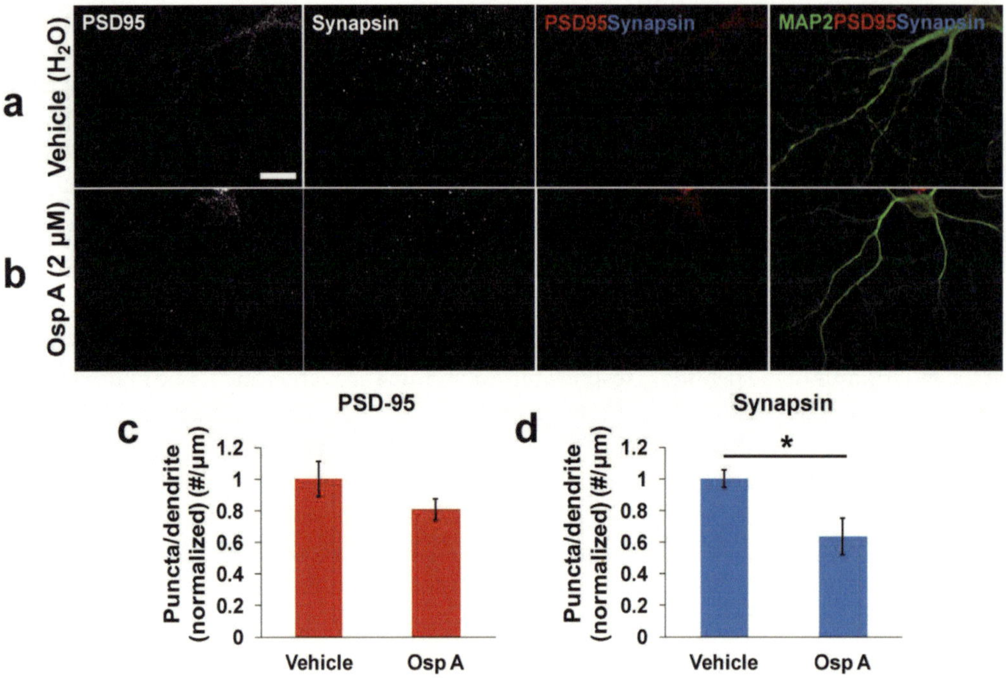

Figure 2. OspA and synaptic density. (a) Fluorescent images showing rat primary hippocampal neurons treated with vehicle (endotoxin-free water). (b) Fluorescent images showing rat primary hippocampal neurons treated with ospA (2 μM). (a) and (b) Neurons stained with anti-PSD-95 (postsynaptic protein), anti-synapsin (presynaptic protein), merge of PSD-95 and synapsin, merge of PSD-95, synapsin, and MAP2 (dendrite) from left to right are shown. (c) The postsynaptic density was measured by the number of postsynaptic sites (puncta) per length of selected dendrite. The postsynaptic density was not affected by ospA ($P > 0.05$). (d) The presynaptic density was measured by the number of presynaptic sites (puncta) per length of selected dendrite. The presynaptic density decreased significantly by ospA ($P = 0.041$).

3.3. Mechanism of synaptic dysfunction induced by bacterial lipoproteins

A recent study demonstrated that viral infection leads to cognitive dysfunction by microglial engulfment of presynapses via the complement C3 pathway [53]. Another recent study showed that viral infection impairs synaptic function via glycogen synthase kinase 3 (GSK-3) activation and intracellular accumulation of Aβ [54]. Thus, an increasing number of studies are being reported that elucidate the mechanism underlying synaptic dysfunction induced by viral infection. Although there is evidence that bacterial lipoprotein ospA also damages presynapses (**Figure 2**), information as to how bacterial infection impairs synaptic function is lacking. Three possible mechanisms may account for synaptic dysfunction during bacterial infection. First, bacterial lipoproteins damage synapses via activation of inflammatory pathways (e.g., TLR2 and TLR4) as discussed in Section 3.1. Second, bacterial lipoproteins damage synapses through neurotransmitter-mediated excitotoxicity. It has been demonstrated that the level of quinolinic acid, the N-methyl-D-aspartate (NMDA) receptor agonist, was elevated significantly in the CSF of Lyme neuroborreliosis patients [55]. The

http://dx.doi.org/10.5772/67053

NMDA receptor mediates synaptic transmission, plasticity, and excitotoxicity in the central nervous system (CNS) and it exhibits excitotoxic effects when an excessive flux of calcium occurs by the increase of a neurotransmitter such as glutamate [56]. However, it is yet to be determined whether the presence of bacterial lipoproteins directly mediates the elevation of quinolinic acid. Third, bacterial lipoproteins damage synapses through physical interaction with synapses independent of biochemical pathways (i.e., inflammation and receptor activation). It has been suggested that the physical properties of proteins (e.g., aggregate pattern and size) is a crucial determinant in mediating pathogenic toxicity [57, 58]. This toxicity occurs independent of their sequences or lengths [59] in a manner that is similar to the aggregation of Aβ in Alzheimer's disease [60] or α-synuclein in Parkinson's disease [61]. Previous studies showed that Pam3-Cys, the synthetic N-terminus of ospA, self-assembled and showed aggregating potential *in vitro* assays [58, 62], which can be related to brain tissue damage including the disruption of synaptic function.

4. Conclusions

This chapter describes the new roles of apoE4 and ospA as major pathogenic endogenous and exogenous lipoproteins, respectively, in neuronal outgrowth and function by discussing recent experimental data in the context of previous reports. Recent studies show that apoE4 enhances neuronal adhesion and axonal outgrowth *in vitro* when it acts alone without lipids. New studies also demonstrate the possibility that ospA can induce synaptic dysfunction by damaging exclusively presynaptic sites. These results contribute to a new understanding of how lipoproteins are involved in developing neuropathology by interacting with neurons. Future studies should focus on the specific mechanism of interaction between apoE4 and neurons and the effect of ospA on synaptic function using *in vivo* models. Along with many pathogenic pathways governed by various cell types in the brain (e.g., microglia, astrocytes, and oligodendrocytes), the effect of pathogenic factors on neuronal activity provides a deeper understanding of structural and functional abnormality in neurodegeneration and neuroinflammation [63]. Understanding the interaction between lipoproteins and neurons in the brain should yield new approaches to the treatment of brain injuries and brain disorders.

Author details

Kwang-Min Kim[1] and G. Tayhas R. Palmore[2]*

*Address all correspondence to: tayhas_palmore@brown.edu

1 Department of Neurosurgery, Stanford University, Stanford, CA, USA

2 School of Engineering, Brown University, Providence, RI, USA

References

[1] Urosevic N, Martins RN. Infection and Alzheimer's disease: the APOE epsilon 4 connection and lipid metabolism. J Alzheimers Dis. 2008 May;13(4):421–35. PubMed PMID: WOS:000256230100007. English.

[2] Wang H, Eckel RH. What are lipoproteins doing in the brain?. Trends Endocrin Met. 2014 Jan;25(1):8–14. PubMed PMID: WOS:000330606100005. English. DOI:10.1016/j.tem.2013.10.003

[3] Latov N, Wu AT, Chin RL, Sander HW, Alaedini A, Brannagan TH. Neuropathy and cognitive impairment following vaccination with the OspA protein of *Borrelia burgdorferi*. J Peripher Nerv Syst. 2004 Sep;9(3):165–7. PubMed PMID: WOS:000224002700007. English. DOI:10.1111/j.1085-9489.2004.09306.x

[4] Pulzova L, Kovac A, Mucha R, Mlynarcik P, Bencurova E, Madar M, et al. OspA-CD40 dyad: ligand-receptor interaction in the translocation of neuroinvasive Borrelia across the blood-brain barrier. Sci Rep. 2011 Sep 8; 1: 86. PubMed PMID: WOS:000296053800002. English. DOI:10.1038/Srep00086

[5] Mahley RW. Apolipoprotein-E - cholesterol transport protein with expanding role in cell biology. Science. 1988 Apr 29;240(4852):622–30. PubMed PMID: WOS:A1988N126500025. English. DOI:10.1126/science.3283935

[6] Uchihara T, Duyckaerts C, He Y, Kobayashi K, Seilhean D, Amouyel P, et al. Apoe immunoreactivity and microglial cells in Alzheimers-disease brain. Neurosci Lett. 1995 Jul 28;195(1):5–8. PubMed PMID: WOS:A1995RM14300002. English. DOI:10.1016/0304-3940(95)11763-M

[7] Hoffmann O, Braun JS, Becker D, Halle A, Freyer D, Dagand E, et al. TLR2 mediates neuroinflammation and neuronal damage. J Immunol. 2007 May 15;178(10):6476–81. PubMed PMID: WOS:000246286200055. English.

[8] Kovacs-Simon A, Titball RW, Michell SL. Lipoproteins of bacterial pathogens. Infect Immun. 2011 Feb;79(2):548–61. PubMed PMID: WOS:000286462000001. English. DOI:10.1128/Iai.00682-10

[9] Mahley RW, Rall SC. Apolipoprotein E: far more than a lipid transport protein. Annu Rev Genom Hum G. 2000;1:507–37. PubMed PMID: WOS:000165768900019. English. DOI:10.1146/annurev.genom.1.1.507

[10] Raber J, Wong D, Yu GQ, Buttini M, Mahley RW, Pitas RE, et al. Alzheimer's disease—apolipoprotein E and cognitive performance. Nature. 2000 Mar 23;404(6776):352–4. PubMed PMID: WOS:000086119000038. English. DOI:10.1038/35006165

[11] Bu GJ. Apolipoprotein E and its receptors in Alzheimer's disease: pathways, pathogenesis and therapy. Nat Rev Neurosci. 2009 May;10(5):333–44. PubMed PMID: WOS:000265365300010. English. DOI:10.1038/nrn2620

[12] Barbour AG, Tessier SL, Todd WJ. Lyme-disease spirochetes and ixodid tick spirochetes share a common surface antigenic determinant defined by a monoclonal-antibody. Infect Immun. 1983;41(2):795–804. PubMed PMID: WOS:A1983RA43900050. English.

[13] Burgdorfer W, Barbour AG, Hayes SF, Benach JL, Grunwaldt E, Davis JP. Lyme-disease—a tick-borne spirochetosis. Science. 1982;216(4552):1317–9. PubMed PMID: WOS:A1982NS80100029. English. DOI:10.1126/science.7043737

[14] Pfister HW, Wilske B, Weber K. Lyme borreliosis—basic science and clinical aspects. Lancet. 1994 Apr 23;343(8904):1013–6. PubMed PMID: WOS:A1994NG68200013. English. DOI:10.1016/S0140-6736(94)90130-9

[15] Ittner LM, Gotz J. Amyloid-beta and tau—a toxic pas de deux in Alzheimer's disease. Nat Rev Neurosci. 2011 Feb;12(2):67–72. PubMed PMID: WOS:000286420100001. English. DOI:10.1038/nrn2967

[16] Liu CC, Kanekiyo T, Xu HX, Bu GJ. Apolipoprotein E and Alzheimer disease: risk, mechanisms and therapy. Nat Rev Neurol. 2013 Feb;9(2):106–18. PubMed PMID: WOS:000315250400010. English. DOI:10.1038/nrneurol.2012.263

[17] Lien E, Sellati TJ, Yoshimura A, Flo TH, Rawadi G, Finberg RW, et al. Toll-like receptor 2 functions as a pattern recognition receptor for diverse bacterial products. J Biol Chem. 1999 Nov 19;274(47):33419–25. PubMed PMID: WOS:000083745200039. English. DOI:10.1074/jbc.274.47.33419

[18] Koch S, Donarski N, Goetze K, Kreckel M, Stuerenburg HJ, Buhmann C, et al. Characterization of four lipoprotein classes in human cerebrospinal fluid. J Lipid Res. 2001 Jul;42(7):1143–51. PubMed PMID: WOS:000169784300018. English.

[19] Fitz NF, Cronican AA, Saleem M, Fauq AH, Chapman R, Lefterov I, et al. Abca1 deficiency affects Alzheimer's disease-like phenotype in human ApoE4 but not in ApoE3-targeted replacement mice. J Neurosci. 2012 Sep 19;32(38):13125–36. PubMed PMID: WOS:000309258100016. English. DOI:10.1523/Jneurosci.1937-12.2012

[20] Farrer LA, Cupples LA, Haines JL, Hyman B, Kukull WA, Mayeux R, et al. Effects of age, sex, and ethnicity on the association between apolipoprotein E genotype and Alzheimer disease—a meta-analysis. Jama-J Am Med Assoc. 1997 Oct 22;278(16):1349–56. PubMed PMID: WOS:A1997YA90500037. English. DOI:10.1001/jama.278.16.1349

[21] Mahley RW, Huang YD. Apolipoprotein E sets the stage: response to injury triggers neuropathology. Neuron. 2012 Dec 6;76(5):871–85. PubMed PMID: WOS:000312283400005. English. DOI:10.1016/j.neuron.2012.11.020 (2012).

[22] Dong LM, Wilson C, Wardell MR, Simmons T, Mahley RW, Weisgraber KH, et al. Human apolipoprotein E—role of arginine-61 in mediating the lipoprotein preferences of the E3-isoform and E4-isoform. J Biol Chem. 1994 Sep 2;269(35):22358–65. PubMed PMID: WOS:A1994PE09800065. English.

[23] Sadowski MJ, Pankiewicz J, Scholtzova H, Mehta PD, Prelli F, Quartermain D, et al. Blocking the apolipoprotein E/amyloid-beta interaction as a potential therapeutic

approach for Alzheimer's disease. Proc Natl Acad Sci USA. 2006 Dec 5;103(49):18787–92. PubMed PMID: WOS:000242689800072. English. DOI:10.1073/pnas.0604011103

[24] Kim KM, Vicenty J, Palmore GTR. The potential of apolipoprotein E4 to act as a substrate for primary cultures of hippocampal neurons. Biomaterials. 2013 Apr;34(11):2694–700. PubMed PMID: WOS:000315748200012. English. DOI:10.1016/j.biomaterials.2013.01.012

[25] LuckenbillEdds L. Laminin and the mechanism of neuronal outgrowth. Brain Res Rev. 1997 Feb;23(1–2):1–27. PubMed PMID: WOS:A1997WN18500001. English. DOI:10.1016/S0165-0173(96)00013-6

[26] Wilson C, Wardell MR, Weisgraber KH, Mahley RW, Agard DA. Three-dimensional structure of the LDL receptor-binding domain of human apolipoprotein E. Science. 1991 Jun 28;252(5014):1817–22. PubMed PMID: WOS:A1991FU06100031. English. DOI:10.1126/science.2063194

[27] Blacklow SC. Versatility in ligand recognition by LDL receptor family proteins: advances and frontiers. Curr Opin Struc Biol. 2007 Aug;17(4):419–26. PubMed PMID: WOS:000250423800006. English. DOI:10.1016/j.sbi.2007.08.017

[28] Goos M, Lange P, Hanisch UK, Prinz M, Scheffel J, Bergmann R, et al. Fibronectin is elevated in the cerebrospinal fluid of patients suffering from bacterial meningitis and enhances inflammation caused by bacterial products in primary mouse microglial cell cultures. J Neurochem. 2007 Sep;102(6):2049–60. PubMed PMID: WOS:000249452000031. English. DOI:10.1111/j.1471-4159.2007.04683.x

[29] Veenendaalhilbers JA, Perquin WVM, Hoogland PH, Doornbos L. Basal meningovasculitis and occlusion of the basilar artery in two cases of *Borrelia burgdorferi* infection. Neurology. 1988 Aug;38(8):1317–9. PubMed PMID: WOS:A1988P586300030. English.

[30] Pfister HW, Rupprecht TA. Clinical aspects of neuroborreliosis and post-Lyme disease syndrome in adult patients. Int J Med Microbiol. 2006 May;296:11–6. PubMed PMID: WOS:000238550200003. English. DOI:10.1016/j.ijmm.2005.12.003

[31] Hirschfeld M, Kirschning CJ, Schwandner R, Wesche H, Weis JH, Wooten RM, et al. Cutting edge: inflammatory signaling by *Borrelia burgdorferi* lipoproteins is mediated by toll-like receptor 2. J Immunol. 1999 Sep 1;163(5):2382–6. PubMed PMID: WOS:000082125400007. English.

[32] Batsford S, Dunn J, Mihatsch M. Outer surface lipoproteins of *Borrelia burgdorferi* vary in their ability to induce experimental joint injury. Arthritis Rheum. 2004 Jul;50(7):2360–9. PubMed PMID: WOS:000222820300038. English. DOI:10.1002/art.20337

[33] Bockenstedt LK, Gonzalez DG, Haberman AM, Belperron AA. Spirochete antigens persist near cartilage after murine Lyme borreliosis therapy. J Clin Invest. 2012 Jul;122(7):2652–60. PubMed PMID: WOS:000306044600035. English. DOI:10.1172/JCI58813

[34] Evans ME, Pollack M. Effect of antibiotic class and concentration on the release of lipopolysaccharide from *Escherichia coli*. J Infect Dis. 1993 Jun;167(6):1336–43. PubMed PMID: WOS:A1993LD69400011. English.

[35] Nau R, Wellmer A, Soto A, Koch K, Schneider O, Schmidt H, et al. Rifampin reduces early mortality in experimental *Streptococcus pneumoniae* meningitis. J Infect Dis. 1999 Jun;179(6):1557–60. PubMed PMID: WOS:000080561100034. English. DOI:10.1086/314760

[36] Bottcher T, Gerber J, Wellmer A, Smirnov AV, Fakhrjanali F, Mix E, et al. Rifampin reduces production of reactive oxygen species of cerebrospinal fluid phagocytes and hippocampal neuronal apoptosis in experimental *Streptococcus pneumoniae* meningitis. J Infect Dis. 2000 Jun;181(6):2095–8. PubMed PMID: WOS:000087923900038. English. DOI:10.1086/315518

[37] Akira S, Takeda K. Toll-like receptor signalling. Nat Rev Immunol. 2004 Jul;4(7):499–511. PubMed PMID: WOS:000222435700012. English. DOI:10.1038/nri1391

[38] Moresco EMY, LaVine D, Beutler B. Toll-like receptors. Curr Biol. 2011 Jul 12;21(13):R488–93. PubMed PMID: WOS:000292805400004. English.

[39] O'Neill LAJ, Golenbock D, Bowie AG. The history of Toll-like receptors—redefining innate immunity. Nat Rev Immunol. 2013 Jun;13(6):453–60. PubMed PMID: WOS:000319445300012. English. DOI:10.1038/nri3446

[40] Ek M, Engblom D, Saha S, Blomqvist A, Jakobsson PJ, Ericsson-Dahlstrand A. Inflammatory response—pathway across the blood-brain barrier. Nature. 2001 Mar 22;410(6827):430–1. PubMed PMID: WOS:000167583800030. English. DOI: 10.1038/35068632

[41] Ramesh G, MacLean AG, Philipp MT. Cytokines and chemokines at the crossroads of neuroinflammation, neurodegeneration, and neuropathic pain. Mediat Inflamm. 2013; 2013:480739. PubMed PMID: WOS:000323444900001. English. DOI: 10.1155/2013/480739

[42] Parthasarathy G, Fevrier HB, Philipp MT. Non-viable *Borrelia burgdorferi* induce inflammatory mediators and apoptosis in human oligodendrocytes. Neurosci Lett. 2013 Nov 27;556:200–3. PubMed PMID: WOS:000328520900040. English. Doi:10.1016/j.neulet.2013.10.032

[43] Ramesh G, Alvarez AL, Roberts ED, Dennis VA, Lasater BL, Alvarez X, et al. Pathogenesis of Lyme neuroborreliosis: *Borrelia burgdorferi* lipoproteins induce both proliferation and apoptosis in rhesus monkey astrocytes. Eur J Immunol. 2003 Sep;33(9):2539–50. PubMed PMID: WOS:000185323200020. English. DOI:10.1002/eji.200323872

[44] Barbour AG, Tessier SL, Hayes SF. Variation in a major surface protein of Lyme-disease spirochetes. Infect Immun. 1984;45(1):94–100. PubMed PMID: WOS:A1984SX63000017. English.

[45] Livey I, Gibbs CP, Schuster R, Dorner F. Evidence for lateral transfer and recombination in Ospc variation in Lyme-disease Borrelia. Mol Microbiol. 1995 Oct;18(2):257–

69. PubMed PMID: WOS:A1995TG43300008. English. DOI:10.1111/j.1365-2958.1995.mmi_18020257.x

[46] Erdile LF, Brandt MA, Warakomski DJ, Westrack GJ, Sadziene A, Barbour AG, et al. Role of attached lipid in immunogenicity of *Borrelia burgdorferi* OspA. Infect Immun. 1993 Jan;61(1):81–90. PubMed PMID: WOS:A1993KF63600011. English.

[47] Weis JJ, Ma Y, Erdile LF. Biological-activities of native and recombinant *Borrelia burgdorferi* outer surface protein A: dependence on lipid modification. Infect Immun. 1994 Oct;62(10):4632–6. PubMed PMID: WOS:A1994PH29800073. English.

[48] Aliprantis AO, Yang RB, Mark MR, Suggett S, Devaux B, Radolf JD, et al. Cell activation and apoptosis by bacterial lipoproteins through toll-like receptor-2. Science. 1999 Jul 30;285(5428):736–9. PubMed PMID: WOS:000081765100055. English. DOI:10.1126/science.285.5428.736

[49] Du XN, Fleiss B, Li HF, D'angelo B, Sun YY, Zhu CL, et al. Systemic stimulation of TLR2 impairs neonatal mouse brain development. PLoS One. 2011 May 6;6(5): e19583. PubMed PMID: WOS:000290305600031. English. DOI:10.1371/journal.pone.0019583

[50] Saijo K, Glass CK. Microglial cell origin and phenotypes in health and disease. Nat Rev Immunol. 2011 Nov;11(11):775–87. PubMed PMID: WOS:000296584700015. English. DOI:10.1038/nri3086

[51] Tang SC, Arumugam TV, Xu XR, Cheng AW, Mughal MR, Jo DG, et al. Pivotal role for neuronal Toll-like receptors in ischemic brain injury and functional deficits. Proc Natl Acad Sci USA. 2007 Aug 21;104(34):13798–803. PubMed PMID: WOS:000249064700050. English. DOI:10.1073/pnas.0702553104

[52] Downer EJ, Johnston DGW, Lynch MA. Differential role of Dokl and Dok2 in TLR2-induced inflammatory signaling in glia. Mol Cell Neurosci. 2013 Sep;56:148–58. PubMed PMID: WOS:000325834100015. English. DOI:10.1016/j.mcn.2013.04.007

[53] Vasek MJ, Garber C, Dorsey D, Durrant DM, Bollman B, Soung A, et al. A complement-microglial axis drives synapse loss during virus-induced memory impairment. Nature. 2016 Jun 23;534(7608):538–43. PubMed PMID: WOS:000378270300051. English. DOI:10.1038/nature18283

[54] Piacentini R, Li Puma DD, Ripoli C, Marcocci ME, De Chiara G, Garaci E, et al. Herpes Simplex Virus type-1 infection induces synaptic dysfunction in cultured cortical neurons via GSK-3 activation and intraneuronal amyloid-beta protein accumulation. Sci Rep. 2015 Oct 21;5:15444. PubMed PMID: WOS:000363122200004. English. DOI:10.1038/Srep15444

[55] Halperin JJ, Heyes MP. Neuroactive kynurenines in Lyme borreliosis. Neurology. 1992 Jan;42(1):43–50. PubMed PMID: WOS:A1992HA89500007. English.

[56] Cull-Candy S, Brickley S, Farrant M. NMDA receptor subunits: diversity, development and disease. Curr Opin Neurobiol. 2001 Jun;11(3):327–35. PubMed PMID: WOS:000169286200009. English. DOI:10.1016/S0959-4388(00)00215-4

[57] Bucciantini M, Giannoni E, Chiti F, Baroni F, Formigli L, Zurdo JS, et al. Inherent toxicity of aggregates implies a common mechanism for protein misfolding diseases. Nature. 2002 Apr 4;416(6880):507–11. PubMed PMID: WOS:000174756500035. English. DOI:10.1038/416507a

[58] Schromm AB, Howe J, Ulmer AJ, Wiesmuller KH, Seyberth T, Jung G, et al. Physicochemical and biological analysis of synthetic bacterial lipopeptides—validity of the concept of endotoxic conformation. J Biol Chem. 2007 Apr 13;282(15):11030–7. PubMed PMID: WOS:000245941500023. English. DOI:10.1074/jbc.M700287200

[59] Pastor MT, Kummerer N, Schubert V, Esteras-Chopo A, Dotti CG, de la Paz ML, et al. Amyloid toxicity is independent of polypeptide sequence, length and chirality. J Mol Biol. 2008 Jan 18;375(3):695–707. PubMed PMID: WOS:000253124600010. English. DOI:10.1016/j.jmb.2007.08.012

[60] Ono K, Condron MM, Teplow DB. Structure-neurotoxicity relationships of amyloid beta-protein oligomers. Proc Natl Acad Sci USA. 2009 Sep 1;106(35):14745–50. PubMed PMID: WOS:000269481000005. English. DOI:10.1073/pnas.0905127106

[61] Conway KA, Lee SJ, Rochet JC, Ding TT, Williamson RE, Lansbury PT. Acceleration of oligomerization, not fibrillization, is a shared property of both alpha-synuclein mutations linked to early-onset Parkinson's disease: implications for pathogenesis and therapy. Proc Natl Acad Sci USA. 2000 Jan 18;97(2):571–6. PubMed PMID: WOS:000084873100012. English. DOI:10.1073/pnas.97.2.571

[62] Hamley IW, Kirkham S, Dehsorkhi A, Castelletto V, Reza M, Ruokolainen J. Toll-like receptor agonist lipopeptides self-assemble into distinct nanostructures. Chem Commun. 2014;50(100):15948–51. PubMed PMID: WOS:000345452400028. English. DOI:10.1039/c4cc07511k

[63] Xanthos DN, Sandkuhler J. Neurogenic neuroinflammation: inflammatory CNS reactions in response to neuronal activity. Nat Rev Neurosci. 2014 Jan;15(1):43–53. PubMed PMID: WOS:000329022000010. English. DOI:10.1038/nrn3617